The Project Engineer's Toolkit

Peter F Cranston

Cranston Engineering Limited
STONEHAVEN, ABERDEENSHIRE

Copyright © 2018 by Peter F Cranston.

All rights reserved. No part of this publication may be reproduced, distributed or transmitted in any form or by any means, including photocopying, recording, or other electronic or mechanical methods, without the prior written permission of the publisher, except in the case of brief quotations embodied in critical reviews and certain other noncommercial uses permitted by copyright law. For permission requests, write to the publisher, addressed "Attention: Permissions Coordinator," at the address below.

Peter F Cranston/Cranston Engineering Ltd
6 Fetteresso Terrace
Stonehaven, Aberdeen South, United Kingdom, AB392DS
www.cranstoneng.com

Book Layout ©2017BookDesignTemplates.com

Ordering Information:
Quantity sales. Special discounts are available on quantity purchases by corporations, associations, and others. For details, contact the Book Sales Department" at the address above.

The Project Engineer's Toolkit/ Peter F Cranston. 1st ed.
ISBN 978-1-9160549-0-5

To my lovely wife Gwen

Thank you. Without your support, patience and many cups of tea, I would have never achieved this goal.

Part 1 Soft Skills

ACKNOWLEDGEMENTS X

PREFACE XII

WHAT A PROJECT ENGINEER DOES 19

    A Project Engineer Understands Why the Scope is Needed 20
    Generates the Big Picture 20
    Responsibility for Scope, Schedule, and Costs 20
    Managing Change 20
    Coordinates Engineering Resources 21
    Coordinates Vendors and Subcontractors 21
    A Project Engineer Provides Leadership 22
    Approving Expenditure 23
    Communication 23
    Differences Between Project Engineering and Other Technical Disciplines 24
    Typical Qualifications 24
    Summary 25

MEETING, CHAIRING, AND FACILITATING 26

    Cost of Meetings 27
    The Benefits of Meetings 28
    Five Types of Meeting 29
    Meeting Planning and Preparation 31
    Running the Actual Meeting 32
    Qualities of Successful Meeting Chairs 34
    Techniques for Getting Everyone Participating 34
    Concluding the Meeting 35
    Post-Meeting Evaluation 36
    Meetings Summary 36

TIME MANAGEMENT 37

    Managing the "Chaos of Inputs" 38
    Time Management Methodology 38
    Batching 39
    Weekly Worksheet 40
    Delegation 41
    Meetings 41
    E-Mails 41
    Managing Distractions 41
    Proactive and Reactive 41
    Mobile Phones and Social Media 42
    Time Management Summary 42

DELEGATION 43

    What is Delegation? 43
    Seven Steps to Delegation 45
    Types of Delegation 47

    Evaluating your Workload Percentage Capacities 48
    Benefits of Effective Delegation 49
    Delegation in Practice 49
    Delegation Summary 50

CONFLICT MANAGEMENT 51
    Definition 51
    Why Conflict Should be Managed 52
    Good and Bad Conflict 52
    Reasons for Conflict 53
    Symptoms of Conflict 53
    Characteristics of Low Conflict/High Performance Projects 53
    Cultural Differences 54
    Virtual Team Considerations 54
    Resolution Strategies 55
    Example of Conflict Resolution 56
    Conflict Summary 57

MOTIVATION 58
    Project Structure 58
    Motivation Theory 59
    Motivators 60
    Fairness 61
    Demotivation 62
    An Example of Demotivation 62
    An Example of Motivation 62
    Motivation Summary 63

PRESENTATION SKILLS 64
    Content 64
    Preparation 65
    Use of PowerPoint 66
    Intended Audience 67
    Position/Stance & Delivery 68
    Audience Engagement 68
    Opportunity to Practice 69
    Performance Nerves 69
    An Effective Presentation 70
    Presentation Skills Summary 70

PROJECT PSYCHOLOGY 71
    Organisation Culture 71
    Types of Manager 72
    Leader Roles 73
    Power 74
    Managing Yourself 75
    Motivation and Leadership 76
    Project Psychology Summary 78

STRESS 79
    What is Stress? 79
    Stressors 80
    Three Types of Stress 81
    Managing Stress 83

- Work-Related Stress 83
- Stress in Colleagues 84
- Stressing the Business 85
- Stress Summary 85

Part 2 Core Project Management

THE POSITION STATEMENT 86
- Objective of the Position Statement 86
- Initial Generation of the Position Statement 86
- 1st Draft Review 87
- 2nd Draft Review 87
- Final Document 88
- Position Statement Summary 88

PLANNING 89
- Level 1 Logic and Schedule 90
- The Project Critical Path 91
- Project Engineers "9 Bar Blues" 92
- Project Work Breakdown Structure 93
- Bottom-Up Planning 94
- How to Use the Planning Information 95
- Planning Progress Reports 97
- Plan Risking 100
- Planning Summary 100

ESTIMATING 101
- The Objectives of Estimating 102
- Features of a Good Estimate 102
- Estimating Definitions 102
- Classes (Types) of Estimates 103
- Estimating Inputs 104
- Elements of an Estimate 105
- Estimate Challenges 106
- Exceptions 107
- Change Control 107
- Estimate Risking 107
- Estimate Approval 107
- Estimating Summary 108

COST MANAGEMENT 109
- Objectives of Cost Reporting 109
- Key Definitions 111
- Cost Reporting Cycle 111
- Role of the Cost Engineer 112
- Cost Management Module 112
- Using and Interpreting Cost Information 113
- Cost Management Summary 115

RISK MANAGEMENT 116
- Definitions 116
- The Risk Management Process 117

- QUALITATIVE RISK ANALYSIS 117
- QUANTITATIVE RISK ASSESSMENT 121
- RISK MANAGEMENT SUMMARY 123

SUPPLY CHAIN 124
- PROCUREMENT 124
- THE PURCHASE ORDER 125
- SUBCONTRACTS 126
- FEATURES COMMON TO BOTH PURCHASE ORDERS AND CONTRACTS 128
- PROCUREMENT STATUS REGISTER AND CONTRACT STATUS REGISTER 129
- EXPEDITING 130
- ONSHORE MATERIALS MANAGEMENT 130
- OFFSHORE MATERIALS MANAGEMENT 132
- SUMMARY 133

CHANGE CONTROL 134
- WHY DO WE NEED CHANGE CONTROL 134
- CHANGE CONTROL REFERENCE DOCUMENTS 135
- THE CHANGE PROCESS 136
- TYPES OF CHANGE 137
- CHANGE REQUEST TURN ROUND TIME 137
- PROJECT ENGINEER & PROJECT MANAGER APPROVAL 137
- MANUAL AND ELECTRONIC CHANGE CONTROL SYSTEMS 138
- THE RISKS OF UNAPPROVED WORK 139
- THE AUDIT TRAIL 140
- THE PROJECT CHANGE COORDINATOR 140
- CHANGE CONTROL SUMMARY 140

PROJECT MANAGEMENT ORGANISATIONS AND BODIES OF KNOWLEDGE 141
- BODIES OF KNOWLEDGE 141
- BENEFITS OF MEMBERSHIP 142
- WHICH ORGANISATION TO JOIN? 143
- PROJECT MANAGEMENT QUALIFICATIONS 143
- PMO SUMMARY 145

Part 3 The Engineering Disciplines

INTRODUCTION TO THE ENGINEERING DISCIPLINES 146
- TYPICAL ORGANOGRAMS 147
- ENGINEERING DELIVERABLES AND INTERFACES MANAGEMENT 148
- FORMAL INTERFACE MANAGEMENT PROCEDURE 149
- ENGINEERING DOCUMENTS PROCESS 150
- KANBAN 150
- ENGINEERING DISCIPLINES SUMMARY 151

PROCESS 152
- THE OIL AND GAS PRODUCTION PROCESS 152
- GATED DEVELOPMENT PROCESS 154
- REQUIREMENT FOR FACILITIES 154
- CONCEPT SELECT STUDIES 155
- FEED 158
- DETAIL DESIGN 161

v

  Commissioning 162
  Operations Support 162
  Drawings and Documents Produced by the Process Discipline 163
  Calculations 163
  Process Summary 164

PIPING 165

  Layouts Management 165
  3-D Model Reviews 166
  Piping Design 167
  Design Survey 169
  Routes to Becoming a Piping Designer/Engineer 170
  Piping Summary 171

MECHANICAL 172

  The Package Management Process 174
  Equipment specification 174
  Supplier Identification 175
  Enquiry and Proposal Evaluation 175
  Package Management 176
  Mechanical Engineering Specifics 178
  Studies 179
  Routes to Becoming a Mechanical Engineer 180
  Mechanical Summary 180

STRUCTURAL 181

  Main Platform Components 181
  Installing an Offshore platform 183
  Items Designed by Structural 184
  Structural Engineering Challenges 185
  The Structural Design Process 186
  Structural Concepts Simplified 186
  Structural Activities and Documents Produced 188
  Example of Structural Failure: The Alexander Kielland 190
  Routes to Becoming a Structural Engineer 190
  Structural Summary 190

INSTRUMENTS AND CONTROL 192

  Main Instrumentation Systems 192
  System Architecture 194
  System Specification 195
  Secondary Instrumentation Systems 197
  Field Equipment 198
  Actuated Instrument Valves 199
  Inputs to Instrument Engineering 199
  Factory Acceptance Testing and Site Acceptance Testing 200
  Safety Integrity Level 200
  Hazardous Area & Ingress Protection Ratings 201
  Intrinsically Safe Systems 202
  Key Drawings and Documentation 202
  Routes to Becoming an Instrument Engineer 202
  Instruments Summary 203

ELECTRICAL 204

  Power Generation 205
  Specifying and Designing the Power System 206
  Power Distribution 208
  Electrical Equipment... 210
  Hazardous Area and Ingress Protection Ratings 212
  Key Electrical Calculations 213
  Key Electrical Drawings 213
  Electrical Equipment Package Management 213
  Electrical Safety Construction and Operating Considerations 214
  Electrical Summary 214

TECHNICAL SAFETY 215
  Inherently Safe Design 216
  Safety Critical Systems and Performance Standards 216
  Major Process Modifications... 216
  Equipment Package Evaluations 222
  Non-Process Modifications 223
  Platform Hazardous Area Zone Plots 224
  Fire & Gas Detection 225
  Temporary Refuge Impact Assessment Offshore 225
  Technical Safety Summary... 226
  Routes to Becoming a Technical Safety Engineer 226

HSSE 227
  Health and Hygiene 228
  Safety Leadership and Accountability 229
  Risk Assessment Management 229
  Sub-Contractor Management 230
  Safety Triangle 230
  Proactive Risk Reduction 231
  HSSE Reports 232
  Security 233
  Environmental 233
  HSSE Advisor Responsibilities 234
  Legal Implications 234
  Behavioural Safety: A Significant Recent Change in Safety Management 234
  HSSE Summary. 235

HVAC 236
  HVAC Applications.. 236
  Safety HVAC Systems 237
  Commercial HVAC Systems 239
  Accommodation HVAC Design Example 239
  Commercial Regulations, Codes, and Standards 242
  Safety Regulations, Codes and Standards... 242
  Design and Interface Issues 243
  HVAC Engineering Deliverables 243
  HVAC Equipment 244
  Construction and Commissioning 244
  Routes to Becoming an HVAC Engineer 245
  HVAC Summary 245

METALLURGY.. 246
  Areas Covered in Metallurgy 247

- Materials Selection 247
- Supplier Technical Assurance 248
- Welding 249
- Integrity Management 251
- Failure Investigation 251
- Coatings 252
- Non-Metallics 252
- Cathodic Protection and Dissimilar Metals 252
- Corrosion Under Insulation 252
- Routes to Becoming a Metallurgist 252
- Metallurgy Summary 253

QA/QC 254
- Quality Assurance (QA) 255
- The Project Management Plan and Quality Management System 255
- Quality Assurance Audits 255
- Quality Matrix 256
- Quality Alerts 257
- Action Request and NCR System 257
- Cost of Quality 258
- Documentation QA Responsibilities 258
- QA Summary 259
- Quality Control 259
- QC Interfaces 259
- Quality Control Activities 259
- Surveillance and Witnessing 260
- Inspection Planning 261
- External Audits 261
- Concessions 261
- Project Changes 261
- Quality Control Summary 261

CONSTRUCTION 262
- Early Input into Design 262
- Main Construction Functions 263
- Onshore Construction Team 264
- Offshore Construction Team 264
- The Onshore Construction Process 265
- The Offshore Environment 266
- The Permit to Work System 267
- The Offshore Work Cycle 268
- New Starts 268
- OIM Walkabout 268
- Construction Progress Reporting 268
- Engineering and Site Queries 269
- Construction Summary 269

COMMISSIONING 270
- Definitions 271
- The Commissioning Process 271
- Completions Management Systems 275
- Inputs to the Commissioning Process 275
- Commissioning Preparation Outputs 276
- Onshore Commissioning Activities 277

- Offshore Commissioning 278
- Onshore and Offshore Discipline Engineer Support 279
- Routes to Becoming a Commissioning Engineer 280
- Commissioning Summary 280

GLOSSARY 281

REFERENCES 285

INDEX 288

Acknowledgements

This book could not have been produced without the valued input, constructive feedback and permissions from the following people and organisations. I thank you all for your support.

Editors

Peter Byrne, BSc, PhD
Eur Ing Mark Davison, CEng, CMgr MCMIMIET
Sidney D Johnston, CEng, MIChemE, MBA, BEng

Industry Engineering/Technical Specialists

Dave Adams, BEng, ACIBSE
Gerard Adams
Mark Anderson, CEng, MIET
Danny Cahill
Finlay Caird, Ceng, MSc, BEng
Stuart Coetzer, CEng, MICE
Stuart Coull, BSc
John Davis, BEng
Paul Equi
Malcolm Forbes, BSc
Paul Foy, CEng, MIET, MEng, MSc, BSc, Dip Mgmt
Fraser Grove
Iona McInnes, CEng, MIMMM, MSc
Susan McNeil, CEng, MIET, MEng ,BA, PMQ
Kenneth Nisbet
Suki Pooni, CEng IChemE, MEng
Neil Robertson
Kris Reid, BA, Dip Risk Management, GradIRM
Matt Ramsay, CEng, MICE
Derek Simpson, CQP MCQI MweldI
Ray Stobbard
Tom Tierney, CEng, MIET,
Eur Ing Jonathan Wells, CEng, FWeldI, FIMMM, MSc, BSc, IWE, EWE
Simon Waddington, CEng, FSEng, MSc
Elvis Zemani, MSc, BSc

Organisations

Many thanks for the following organsiations who have kindly supported with permission to use drawings, photographs, figures and sample documents.

- BP Plc.
- CNOOC UK Ltd Petroleum UK Ltd.
- ConocoPhillips UK Ltd.
- Covey Franklin.
- Gexcon UK Ltd.
- Quantum Controls Ltd.
- Repsol Sinopec Resources UK Ltd.
- Siemens AG.
- Whittaker Engineering Ltd.
- WorleyParsons Ltd.
- WOOD-Commissioning.

Preface

This book is the result of 10 plus years of gathering information relating to Oil & Gas Topsides Project Engineering. It is intended to provide readily usable tools and techniques to aid the reader in the performance of his/her responsibilities as a Project Engineer. As the Oil and Gas industry strives to become more effective and more efficient, project engineering is about working ***smarter not harder***.

How to use this book

For time-limited Project professionals, I suggest you review the following mindmaps to determine which chapters might be most beneficial, and then complete them in your priority sequence. I also suggest using a highlighter to mark key sections and make your own notes directly in the book; it is your toolkit!

In addition, there are several suggested actions and exercises to guide you as you work through the book. These are designed to help you during your specific projects.

Book layout

This book is split into 3 main sections:

- Soft skills.
- Core Project Management.
- Engineering Disciplines.

Mindmaps

Mindmaps are principally used at the start of each chapter to summarise the contents of that chapter. They are intended to be "read" in a clockwise manner, starting from the top right side. In developing each chapter, I started out with a detailed mindmap, and I recommend that you consider using mindmaps as well throughout your project engineering work.

Soft skills

Soft skills are vital for a Project Engineer who will spend a significant portion of his/her time dealing with people of various backgrounds, competencies and attitudes. These soft skills can be developed and practised to great effect.

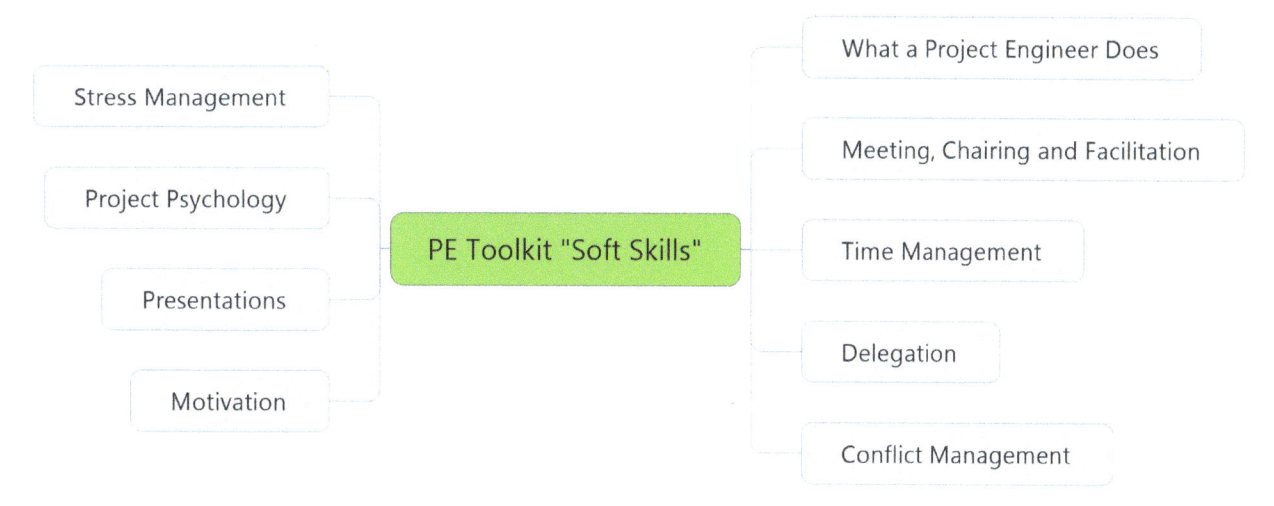

PE Toolkit "Soft Skills"
- Stress Management
- Project Psychology
- Presentations
- Motivation
- What a Project Engineer Does
- Meeting, Chairing and Facilitation
- Time Management
- Delegation
- Conflict Management

Core project management

This section explores the core project management competencies associated with Topsides Project Engineering and uses actual documentation examples to explain how to manage these areas.

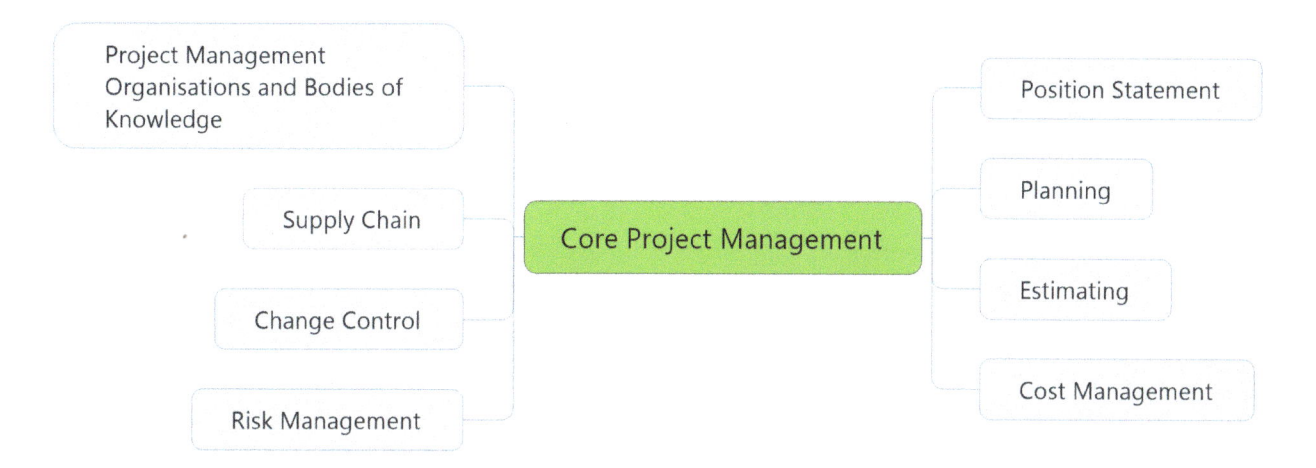

Core Project Management
- Project Management Organisations and Bodies of Knowledge
- Supply Chain
- Change Control
- Risk Management
- Position Statement
- Planning
- Estimating
- Cost Management

The engineering disciplines

This section demystifies the Engineering Disciplines in the Oil and Gas industry, and provides insight from a Project Engineers viewpoint. These chapters will allow you to understand the work processes for each discipline, plus their associated inputs and the outputs so that you will be more effective in integrating all the engineering elements into your project scopes.

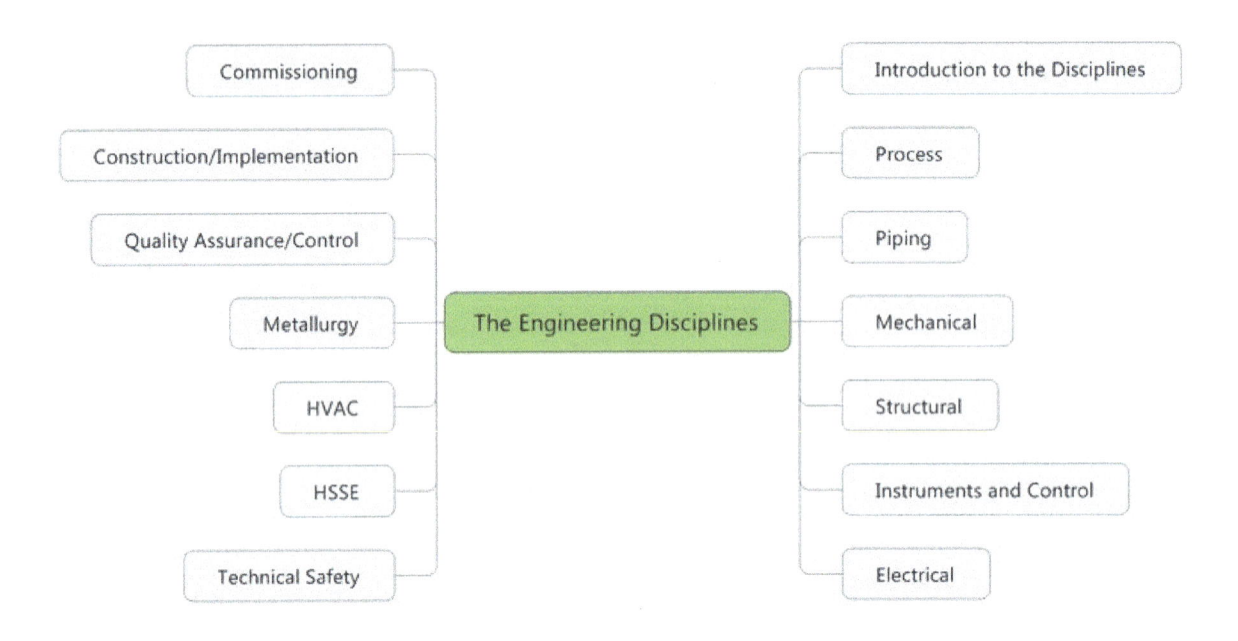

```
                                                            Introduction to the Disciplines
      Commissioning
                                                            Process
      Construction/Implementation
                                                            Piping
      Quality Assurance/Control
                          The Engineering Disciplines       Mechanical
      Metallurgy
                                                            Structural
      HVAC
                                                            Instruments and Control
      HSSE
                                                            Electrical
      Technical Safety
```

Influences on the content

The content has been influenced by the following three main sources:

30+ Years of project engineering/management experience

Having worked as a Project Engineer for over 30 years, I have gained much hard-won experience, which has resulted in developing techniques that work in a "real world" project environment.

Project management body of knowledge

As a member of project management organisations since 1999, including the Association for Project Management (APM) and the Project Management Institute (PMI), my projects were run using the principles contained in project management industry books of knowledge.

Toastmasters International

The skills and techniques in leadership and communication are influenced heavily from over 11 years of active membership in Toastmasters International. Toastmasters is a worldwide organisation of over 350,000 members promoting development of both leadership and communication skills. These techniques have been adapted for the requirements of Project Engineering within oil and gas engineering contractors.

Feedback

Continuous improvement is key to everyone's development; hence I would welcome feedback on what works for you and what does not. These comments will be reviewed and responded to. Please e-mail peterc@cranstoneng.com

Website

In parallel with this book, there are a series of webinars that provide additional materials. Please see **www.cranstoneng.com.** In conjunction with respected industry experts, we are planning additional specialised subject webinars and would welcome your specific subject requests.

I hope you enjoy your journey through the Project Engineers Toolkit and apply the tools where appropriate whilst developing your own individual management style.

Peter Cranston. April 2019

The greatest accolade you can receive as a Project Engineer is a simple, "You deliver!"
—Unknown

CHAPTER 1

What a Project Engineer Does

What is a Project Engineer? It is estimated that there are circa between 100,000 and 200,000 Project Engineers (PEs) in the worldwide Oil and Gas industry, so you are or will be part of a very large community. The term Project Engineer (PE) covers a multitude of areas and varying degrees of technical knowledge, project size and responsibilities, all having the common feature of accountability for the delivery of projects or part of a project.

```
                                    What a Project Engineer Actually Does
  - Anything required to successfully deliver the project          - Understands why scope is needed
  - Ensures clear and concise communication                        - Generates the big picture
  - Approves expenditure within delegated authority limits         - Responsible for scope, schedule and estimate
  - Interfaces with clients representatives                         - Manages change
  - Provides leadership to problem/issue resolution                 - Coordinates engineering resources
                                                                    - Coordinates vendors & subcontractors
```

Figure 1.1 What a PE does Mind Map.

Project Engineers work for Oil & Gas operators, engineering contractors, equipment manufacturing subcontractors and maintenance companies. They have an essential role to play within technical project teams in ensuring that engineering/construction/maintenance projects are completed safely, within budget and in a timely manner.

A Project Engineer may be responsible for managing a project from thousands up to many millions of dollars. They may be part of a project management team responsible for parts of a larger project, or they may be responsible for managing a portfolio of smaller project scopes. The role of the Project Engineer can often be described as that of a liaison between the Project/Contract Manager and those within the technical disciplines relevant to the project. The Project Engineer is also often the primary point of contact for the client.

As a PE, it is standard for you to be responsible for successful delivery, and for managing scope, cost, schedule and risk. The greatest accolade a Project Engineer can receive is a simple two words: *"You deliver!"*

A Project Engineer Understands Why the Scope is Needed

Many projects run into problems because the scope is either poorly defined, poorly understood, or both. The Project Engineer needs to fully understand the scope and why it is needed so he/she can provide the optimum solution using the resources at his/her disposal. A structured technique for defining, agreeing and maintaining scope alignment will be explained in chapter 10, "The Position Statement."

Generates the Big Picture

The Project Engineers needs to set the direction the team will proceed in and supports team members in reaching the goals. Sometimes the individual engineering disciplines/subcontractors are so focused on their individual scopes, that they may not appreciate the bigger picture. By taking time and discussing with both internal and external stakeholders, the Project Engineer can build the "big picture", manage the interfaces, and communicate to the wider team.

Being able to simplify and communicate complex issues and scopes is a vital Project Engineering skill because it allows team members to understand the overall scope, and thus be better able to understand where their contribution fits.

Responsibility for Scope, Schedule, and Costs

In most projects, the Project Engineer is responsible for scope cost and schedule, although this can vary on the specific project.

- Scope is what we are going to do.
- Cost is normally the budget of the project.
- Schedule is when we are going to do it.

Managing Change

Change is in the nature of projects, be it scope, schedule, cost or other influences. Badly managed change can result in cost and schedule overruns and an unhappy client. However properly managed change ensures that there are no surprises and that changes are agreed before implementation.

On many projects, the Project Engineer is responsible for managing change requests/variations and for submitting them to clients for approval. Some lucky Project Engineers may have a Change Coordinator to assist them with the workload associated with the changes. The job of Change Coordinator is an excellent position for new recruits/graduates as it exposes them to all areas of the project and reinforces the importance of change control from day one.

Poor management of change is a common reason for a "bad" project. Change is integral to all projects and properly identifying and managing change is an important skill for the Project Engineer.

Coordinates Engineering Resources

The Project Engineer will be responsible for coordinating engineers from various disciplines including Process, Structural, Piping, Instruments, and Electrical, to name but a few. I have specifically used the word "coordinate" because in many projects, the Discipline Engineers report to their discipline leads for work allocation and technical guidance, although their efforts are coordinated by the Project Engineer.

The simplest way to view coordination is that the Project Engineer represents the client's interests while simultaneously being a customer for the Discipline Engineers.

Coordinates Vendors and Subcontractors

Almost all projects will require the services of subcontractors and vendors.

Subcontractors

Subcontractors are suppliers that enter into specific work contracts with the main contractor to work on major projects or for companies that need specific tasks to be completed within a limited period. Companies usually enter into subcontracting arrangements because they do not have the expertise in-house.

Vendors

Vendors are suppliers that sell identical or similar products or services to different customers as part of their regular operations. Examples include procurement of catalogue parts and components.

Project Engineers are often responsible for coordinating and integrating subcontractor and vendor services into the overall project delivery. Many PEs are employed by subcontractors and vendors responsible for their company's contribution to the greater overall project.

Typical organisations

Project Managers and Project Engineers can exist at all levels in a project as illustrated below. Defined roles and responsibilities are more important than job titles.

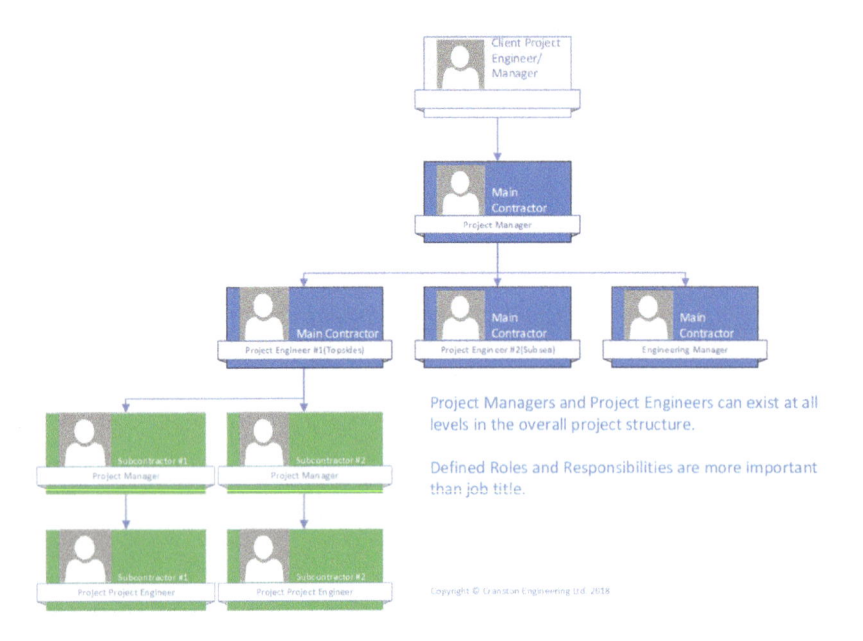

Figure 1.2 Typical Project Management Organisation.

A Project Engineer Provides Leadership

One of the most important traits of a good Project Engineer is excellent leadership skills. However, do not worry if you do not currently rate yourself as a strong leader since leadership skills can be learned and practised. Most people already have some leadership skills, and in this book we will look at techniques to develop those skills.

Resolution of issues.

In projects, many issues are not black and white and will call for some project support to resolve. The Project Engineer will be able to facilitate resolution of multi-discipline and multi-company issues. Guidance, support, and coaching in resolution will normally be provided by the Project Manager or other more senior Project Engineers.

The key to resolution is to fully define the issues, communicate to all the required people, and work systematically to an agreed upon and documented solution.

Defusing issues

As your experience develops you will be able to anticipate problems and defuse them before they impact the project. To outsiders, some of the most successful Project Engineers seem to be calmly sailing along without any issues, but underneath the surface they are working hard to identify and prevent problems from building up. This should be your aspiration.

Making decisions

Often you have to facilitate a decision to progress a project scope, and you may have to do this based only on partial information. Depending on your project, you may work through an issue with the team, document it, and having the bigger picture, `make a recommendation to be ratified by your manager. All project decisions should be recorded in the project decision register including at a minimum, the decision, date of decision, decision parties, and any key document attachments. Making difficult decisions will develop good judgment; you might not always get it right, but when you don't, you will certainly learn for the next time.

Client interface

The Project Engineer is normally the direct single point of contact with the client's representative for the project or part of a project. This will require regular verbal and written communication. Your people management skills will be developed and tested as you work with clients ranging from pleasant, challenging but fair to downright rude. The key is to remain polite and professional at all times since you are the face of your organisation.

Approving Expenditure

As a Project Engineer you will normally be responsible for elements of the budget and for approving expenditure up to a certain authorised level. You will need to be able to challenge the contents of an estimate when required to ensure that both your organisation and your client are getting value for money.

Communication

When asked, "What are the three most important aspects of Project Engineering?, my answer is: communication, communication, communication. You cannot over communicate. In the fast-moving pace of a project, miscommunication means errors, schedule slippage, and cost increases.

By good communication, I do not mean endless e-mail trails and turgid texts, but rather simple, concise communication that provides the right information in the right format to the right people. People are busy, so time spent on producing concise and accurate information will get results.

Communication Methods

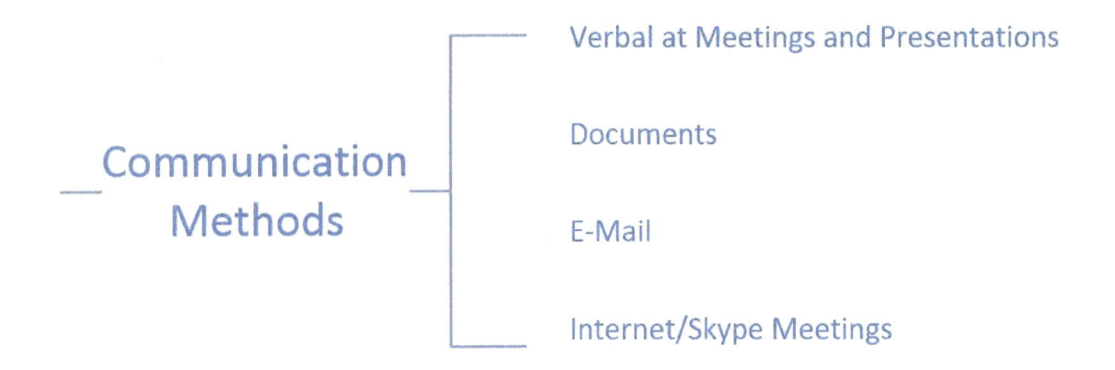

- Verbal at Meetings and Presentations
- Documents
- E-Mail
- Internet/Skype Meetings

Figure 1.3 Main Communication methods.

Less is more

When people are busy and swamped with data, take time to distil your message to the relevant points and actions and send only to the required people. This allows for more information to be received. ***Less is indeed more.***

Differences Between Project Engineering and Other Technical Disciplines

A Project Engineer will not normally complete the detailed engineering deliverables, but will be responsible for coordinating and managing the efforts of the technical disciplines and other specialists. To do this effectively you will have to develop and enhance your people skills. You will not need to get involved in every last detail and instead will sift out the important items from the "chaotic noise" to assist the team in delivering.

You need to be delivery-focused and able to self-manage. To be able to lead and manage a team, you first have to be able to manage yourself.

Typical Qualifications

There are three main routes for Project Engineer entry:

- Academic.
- Project Management.
- Experience-Based.

Qualifications

Although the education and training possessed by Project Engineers varies by field, most hold a degree in engineering, construction management, or business management. Some Project Engineers also choose to earn a bachelor's degree or a Master's degree in Project Management. There are many excellent Project Engineers who had minimal qualifications but came up through the ranks.

Project engineering/management professional qualifications

Within Project Engineering/Management, there are several International Project Management organisations that are worth considering joining to gain certifications by demonstrating experiences and passing exams. These certifications provide common Project Management language and industry-accepted levels of competency.

Project management organisations include:

- Project Management Institute   www.pmi.org.
- Association for Project Management   www.apm.org.uk.
- International Project Management Association   www.ipma.world.

Your employer may already have corporate membership with one of these organisations.

Personal attributes

Project Engineers need to have to have something more than just intelligence and training. They need to have the right personal attributes, which includes:

- Good problem-solving skills.
- Logical structured thinking.
- Dogmatism.
- Good people skills.
- Being cool under pressure.
- And above all, a determination to get the job done.

Summary

As you have now seen, Project Engineers do a lot and are normally very busy people. If you like a multifaceted challenge, then Project Engineering may be right for you. Armed with the basic toolkit in this book, you will be better able to face the numerous challenges that are part of a Project Engineer's daily work life. As your experience develops, you will be able to add additional tools and techniques to your toolkit.

By continuously putting yourself slightly out of your comfort zone, you will develop in Project Engineering/Management skills. Over the years, I have worked with many Discipline Engineers who have made the move into project management. The majority of them like the additional breadth, challenges, and increased satisfaction, and do not wish to return to their disciplines.

The most important attribute for a budding PE is a positive delivery-focused "can do" attitude.

Some say that "Project Engineers oil the wheels of a project."

I personally like to think that Project Engineers do "whatever needs to be done for the project to succeed."

Meeting, Chairing, and Facilitating

Meetings occupy a significant proportion of Project Engineers' time. I regularly run a meetings effectiveness workshop and usually start by asking the attendees, "Is there anyone who could not have more efficient meetings?" Without exception the answer is always *"Yes"*.

Figure 2.1 Meetings Mind Map.

I then ask if they have ever attended meetings where people turn up late or there is no written agenda. There is agreement once again. You may notice, possibly with a wry smile, threads of similarity, from the following imaginary meeting:

People arrive late or not at all. The meeting starts late. There is no pre-issued agenda. There is tremendous discussion and argument. The meeting overruns and people leave having voiced their opinions. The meeting closes with no minutes or action logs, and is repeated all over again the following week.

Features of a "Bad" Meeting

- People Arrive Late or Not At All
- Meeting Starts Late
- No Pre-Issued Agenda
- Little Participant Preparation
- Meeting Overruns
- No Minutes or Action Log

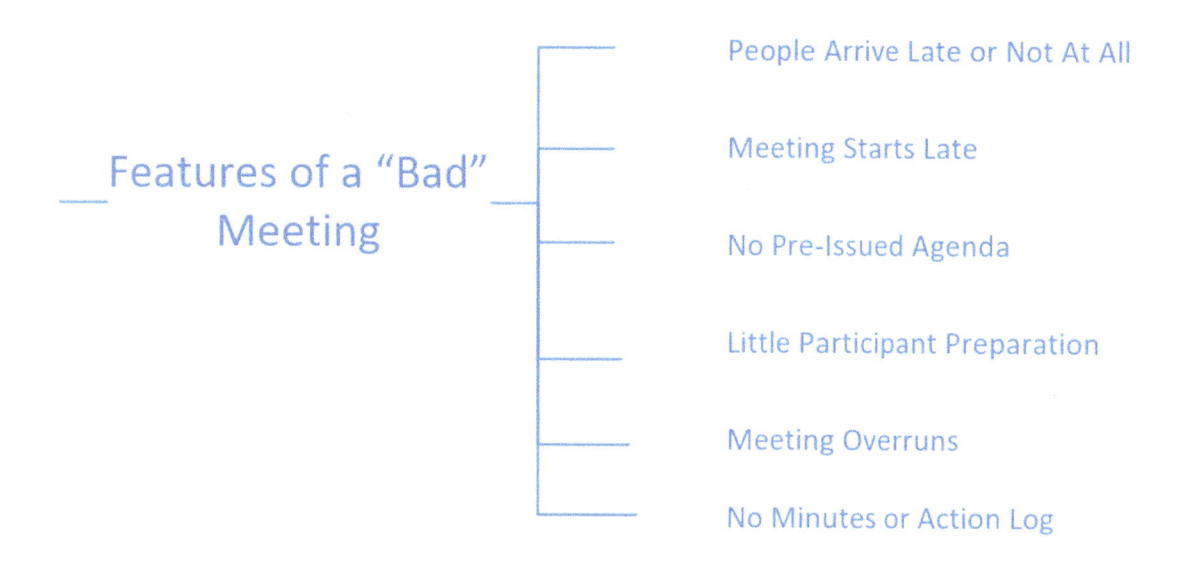

Figure 2.2 Features of a "bad" meeting.

I call this type of meeting "A chat round the campfire." It's great fun, but is it effective?

This chapter will provide the basic tools to allow your meetings to improve in quality and effectiveness, which will benefit the entire project. It is also one of the longest sections, which reflects its importance.

Cost of Meetings

It is not uncommon for Project Engineers to spend over 70% of their time in meetings. At meeting workshops, one of the tasks I ask attendees to do is to make an estimate of the cost to the company of a fairly large meeting with 20 attendees for an hour. Try this yourself by multiplying 20 attendees by the organisation's average charge-out rate, including overhead for the team members. For example 20 engineers at a rate of £65/hr equating to £1,300

As a project leader, it is incumbent on you to spend the company and client's money wisely, so it is important to make your meetings as effective as possible.

The Benefits of Meetings

Well-prepared and well-run meeting meetings are essential to the success of the project because they are a vital communications tool. Carried out properly, they have many positive results:

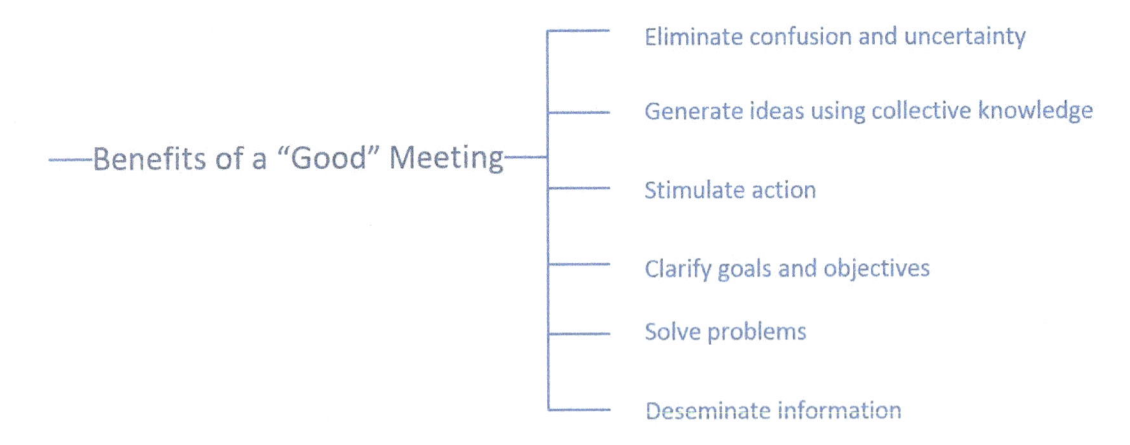

Benefits of a "Good" Meeting
- Eliminate confusion and uncertainty
- Generate ideas using collective knowledge
- Stimulate action
- Clarify goals and objectives
- Solve problems
- Deseminate information

Figure 2.3 Benefits of a "Good" meeting.

Coordination of team members

Trying to individually coordinate all your team members is not only almost impossible, but also very time consuming, and you cannot actually have instant interface, communication, and feedback. A meeting can be a very efficient use of your time to discuss issues and get agreement on actions/action parties.

Generate ideas

Meetings are great for generating ideas since team members can build on one another's contributions. The sum of group output will always be greater than the sum of the individuals alone.

Solve problems

Meetings can be used to solve problems using the five steps shown in Figure 2.4, page 29.

Generate agreement of actions

Meetings are great for agreeing on actions and action parties. If you have the required team or sub-team at a meeting it is usually clear who will complete the action. There is also a gentle peer pressure to complete the agreed upon actions by the required dates as no one likes to be the one who lets the team down.

Provide clarification

Although the Project Engineer should be clear on the objectives, most of the team members will come to the meeting with slightly, or even totally different understandings of the background, issues and required actions. This is your opportunity to get everyone on the same page.

Tip: Clear, concise presentations with clear visuals are good for aligning the teams understanding. Try to avoid too much text.

Your meetings will give you and other members of the team the opportunity to communicate the objectives and requirements of the particular topic of the meeting.

To disseminate information

People are constantly bombarded with e-mails where key information is lost in the noise.

Tip: During a normal week or month of a project, there are many pieces of information and issues that need to be brought to the team's attention. An effective way to do this as PE is to trawl through you e-mails/memos etc. and decide which items need to be highlighted for the meeting attendees. Print these out with any supporting notes and then systematically work through this brief at the meeting.

Formal record of agreements

It is vital that records are kept of meetings, especially with clients and subcontractors to cite variation requests and claims. This not only protects your clients, but yourself and your project team.

Five Types of Meeting

The first step towards planning a meeting is defining what type of meeting it is. While every meeting is unique, being familiar with the five most common types of meetings will help you better identify the goals, structure, and activities best suited for a particular meeting.

- Status update meetings.
- Information sharing meetings.
- Decision-making meetings.
- Problem-solving meetings.
- Team-building meetings

Status update meetings

These can also be known as planning meetings or project update meetings. Typically, the Project Engineer is assisted by another team member, and will run through action logs and plans seeking updates on status, progress, issues and next steps from the team members. A key feature of this type of meeting is that resolution of any problems and issues will normally be taken outside the meeting, with the focus in the meeting being on status updates.

Information sharing meetings

This is where the Project Engineer, the client or other team members share information with the attendees, who are normally passive listeners that you can validate understanding with and, when appropriate, request feedback.

Decision making meetings

At these meetings, decisions are formally recorded in the minutes and, for significant decisions or changes, in the project decision register. It is important that decisions are formally recorded so that all parties can later reference the information, if need be.

Problem-solving meetings

In this type of meeting, all invited team members are requested to contribute in resolving the issue. The role of the Project Engineer is to facilitate problem-solving and maintain the momentum, ensuring discussions are relevant. These meetings should allow exploration of issues without getting too far off track; a challenging task for the PE that takes skill and practice to perform effectively.

To solve a problem, I suggest you follow the five steps below:

Outline Problem or Issue → Analyse Causes of Problem or Issue → Suggest Solutions → Evaluate Solutions → Select and Implement the Best Solution

Figure 2.4 Problem-solving steps.

Team building meetings

Team building meetings can either be regular or one-off events.

One-off events

For one-off events, there will usually be a specific project task to be completed, such as scope definition, lessons learned, or plan development, each of which may take a half or even a full day. These meetings will often use an experienced external facilitator.

The typical format for one-off meetings usually start off with selecting an external venue, with the meeting starting with breakfast. "Ice breaker" introductions are good for getting people to relax. Breaking the team into smaller groups for each of the day's activities allows members to interact with one another and become more comfortable working together. Team-building meetings such as these often finish with an evening meal or perhaps even drinks at the bar, as appropriate to the culture.

Regular team meetings

A regular weekly meeting with the team is also a good team-building opportunity. On a recent project, we implemented a Friday "wash-up" meeting just before lunch. Since it was a Friday, everyone was more relaxed and looking forward to the weekend. Following a short cascade of information, each team member was invited to brief on what they were doing, request any support they required, and offer support. Light-heartedness was also encouraged to promote the forming of a strong supportive team.

"Well-run meetings engender client confidence."

Meeting Planning and Preparation

There are many meetings of the type I call, "The chat round the campfire" meetings. These are unprepared, unstructured, and unproductive. They have the following features: no preparation, unclear objectives, loose discussions, no formal agreed-upon actions or action parties and, in my view, little value.

Preparation time

One of my simple rules of thumb for meeting preparation for a typical one-hour meeting is to dedicate one-hour for planning and one-hour for follow up. By following these simple steps your meeting will be more effective, efficient, and satisfying.

Carefully select the attendees

Only invite attendees who really need to be there, either from a knowledge-input viewpoint or as a decision maker. Who to invite is critical in ensuring an effective meeting:

- Who is knowledgeable on the subject?
- Who has influence?
- Who can help?

Sometimes you must know who *not* to invite when their involvement might be counter-productive or even destructive to the meeting objectives.

If you can, always appoint a minute-taker to record the agreed actions since trying to chair and take minutes is difficult. Pre-meetings can be used for great effect to warm up various parties, such as for seeking support or clarification prior to the main meeting. Finally, be aware that having additional non-contributing attendees will dilute the effectiveness of the meeting.

Select an appropriate location

If you can impact this, try to set the meeting in surroundings that are attractive and comfortable: Not too hot or cold, and of a suitable size for the number of attendees. A room that is too small will probably be stuffier, while a room that is too large may be draughty and intimidating, and long, narrow rooms can impede discussions. A big meeting may require an external facility, such as a meeting room in a local hotel. For giving information to a large group. a classroom format is good. For discussions, it is best to set the room for attendees to face each other.

Style of meeting

As you will be chairing the meeting, you should consider the style of meeting and the atmosphere you wish to create.

I used to run design review meetings in which only the Discipline Engineers were invited. I picked Friday mornings since the dress was casual and the atmosphere more informal. We had breakfast rolls and coffee, and the discussion focused purely on the technical aspects of the project. This allowed them to open up and discuss some potentially sensitive/major issues.

In contrast, a typical project progress weekly meeting is a much more formal event. We look at performance and costs and in-depth discussions are discouraged, but kept for other meetings. Progress meetings are typically held earlier in the working week.

Inform the attendees in good time

Many meetings are called with little or no notice. Those who attend the meeting are likely irritated and potentially uncooperative. Participants should be invited days, if not weeks in advance, depending on their availability and the importance of the meeting. When possible, give attendees as much notice as possible to allow them to sort out their calendars. Provide any pre-reading beforehand as well so that attendees are briefed before they attend the meeting. Advise any key meeting members of any preparation that you require of them to complete prior to the meeting.

I normally use Microsoft Outlook for meeting invites, as it can provide time, location and supporting information at the click of a mouse, and has a feature for Skype. There will be times when meetings must be called on very short notice, but these should be the exception rather than the norm in a well-run project.

Develop and issue an agenda

Issue an agenda with the meeting invite in order to identify who you wish to support each agenda item. If required, issue a draft of the agenda to key attendees even earlier to allow them to include any items they wish to discuss. Appoint a minute-taker prior to the meeting as it is difficult to both chair and take minutes.

The agenda, notes, and any background readings should be issued as early as possible to allow time for gaining familiarity with the meeting topic and purpose. Include the following when sharing information about the meeting:

- Meeting date, time and place.
- Purpose and outcome required.
- Main topic headings.
- Information attendees are requested to read beforehand, bring with them.
- What is expected, presentations, reports etc.
- Meeting length and any domestic arrangements, i.e. breaks, meals etc.

Running the Actual Meeting

The main thing to remember is that you are the chair of the meeting and that you have control.

Arrive early and kick-off the meeting on time

You and your minute-taker should arrive early to ensure the room layout is suitable and any audio-visual tools are working properly, with IT being notified to address any potential issues before the meeting starts. For a regular meeting, being 10 minutes early should suffice. For a large, external meeting in a hotel, one hour would be more appropriate.

As chair, it is your responsibility to start the meeting on time and close the door to embarrass late arrivals. If we cannot start meetings on time, what does it say about our overall Project Management? Set a personal target of starting on time and finishing on time. Everyone is busy.

The Chair/Project Engineer is the driver for the meeting. Chairing meetings can be hard work but with the correct preparation, the meetings will be efficient, productive and satisfying for all attendees. The meeting chair sets the tone, the objectives, and the agenda and steers the meeting within the allocated time.

Welcome

Welcome everyone and if there are any new attendees, go around the room with introductions. This will usually include everyone's name and role on the project or job. Circulate a "Sign in" sheet for new client or subcontractor meetings. You can later scan this and add it to the project file as a record of attendees. Some attendees may be nervous in meetings, and it is up to the chair to put them at ease. Chat informally if appropriate, and tell a short story or anecdote lighten things up.

Set the scene

Brief everyone on the background and objectives of the meeting and then get into the detailed agenda. If the meeting is to discuss a new issue or problem:

- Explain how the situation arose.
- State why the problem or issue is important.
- Ask participants how the problem affects them.
- Point out how each participant can contribute to a solution.
- Explain the group's responsibility for solving the problem.

Stick to the agenda

The chair should keep the meeting on-track and follow the agenda. The chair should regularly summarise and ask for actions to be recorded.

Sharing the workload

For regular project or client meetings, share out the responsibilities. This ensures more people have preparation responsibilities and reduces your overall workload.

Handling challenging personalities

This is the skillful part that comes with practice. The group may digress or become stuck on one topic. Here you will find the following:

- The derailer (derails for spite/ulterior objectives).
- The arguer.
- The interrupter.
- The silent but knowledgeable one.
- The talker. Why use one word when you can use 10?
- The one who monologues.

You have probably seen all these behaviours before. Dealing with difficult people at meetings can be of the most stressful and challenging tasks for the meeting chair, but here are a few tried and tested tips:

Tips for dealing with 'big personalities'

- Prepare. Arrive early and take the best, most central, seat in the room.
- Keep to the agenda and remind people of limited time at the start of the meeting.
- Stand up and thank the person for input and suggest that others need to contribute.
- Put contentious issues in a "parking lot".
- Warm up 'personalities' prior to the meeting that might inhibit the success of the meeting.
- Practice and develop your own techniques, i.e. gravitas, humour, and bringing in other contributors.
- Post-meeting, have a one=-on-one conversation with the problem attendees, advising them on their problem behaviours and seeking agreement for modification of said behaviour. If unsuccessful, raise the issue with your manager and the problem attendee's manager.

Qualities of Successful Meeting Chairs

Good meeting chairs will seek to develop the following qualities:

- Poised, Confident, and Prepared: Try to stand or sit straight and establish eye contact with the attendees.
- Sensitivity: To other people and their strengths and limitations.
- Impartiality: Remain impartial until point of decision-making.
- Tact: Able to say the right thing in the right way at the right time.
- Good Judgement: Which will develop with practice.
- Good Listening Skills: This is one of most important skills to develop.

Techniques for Getting Everyone Participating

Sometimes getting people talking and discussing can be a challenge. The following steps can be used to get things moving in the right direction:

- State that you need ideas from the assembled group.
- Ask open questions
    - "What do you think?"
    - "How do you think we can achieve this goal?"
- Invite cooperation.
- Seek the unique knowledge of attendees.
- Call on someone directly: "Michael, you have experience in this subject, what do you think?"
- Give compliments to individuals.

Promote an open atmosphere

Be sincere in your desire for input. Use appropriate body language, and avoid evaluating or judging early in the discussion; this can happen later.

Summarise

Regularly summarise what has been achieved to keep focus on the issue or issues. This keeps things moving forward.

Use transitions

"We seem to have exhausted this item, we will now move on to the next point." You are the leader. You have control. All it takes is practice.

Test possible solutions

Summarise and review the solutions and seek the group's opinions on them. Recording these on a whiteboard or flipchart helps focus the group.

Keep the discussion on track

This requires skill. Knowing when to interrupt, change track, or let the discussion flow is difficult to judge. Practice is the key, and don't forget to review how a meeting went afterwards, reflecting either on your own observations or by seeking feedback from a trusted team member. Even after 30 years, I am still learning.

Concluding the Meeting

Ensure you conclude on a positive note, even if the objective has not been achieved.

Six steps to conclude:

- Indicate that it is time to conclude, i.e., "We've got 5 minutes left, so I will summarise and review the agreed actions".
- Review the original problem or issue.
- Summarise the progress made.
- Emphasise the agreements and decisions made.
- Run through the agreed actions, action parties and dates.

Tip: Task only one person to an action item, and they must have attended the meeting. Two people assigned to one task will reduce the likelihood of the task being completed, and with three there will be no chance!

Tip: Advise that the meeting minutes or action log will be issued to the attendees in an agreed upon distribution method.

Tip: Detailed minutes are good for kick-off meetings and contractual meetings. In a commercial claim situation, clear agreed-upon minutes can save the company from significant claim costs.

Tip: For regular internal meetings and where trust has developed you can use action logs.

If there are no written actions, then why have the meeting? At a minimum, there should be a short e-mail with a summary and agreed actions/action parties. There are some exceptions, such as sensitive discussions of a performance or personal nature.

Tip: Issue the minutes or action log within a day if possible, and advise the date for the next meeting, if required.
Tip: Finally, thank the group for their input.

Post-Meeting Evaluation

I recommend that you evaluate your meeting quality, especially in the early part of your career as a Project Engineer. There are three sources you can use:

Yourself: What went well? What went not so well? What would I change?
An Observer: Ask a trusted team member whose opinion you respect.
Participants: Especially for a workshop, issue a short anonymous questionnaire.

Meetings Summary

Project Engineers are key leaders in projects and can significantly influence success. You now have the meeting management tools. Push your limits and try them out. Aim for continuous incremental improvement. Remember, practice makes perfect. I myself am still practicing and learning.

"The majority of meetings should be discussions that lead to decisions."
- Patrick Lencioni

CHAPTER 3

Time Management

We have all seen colleagues, clients and even ourselves juggling meetings, replying to e-mails in meetings, and taking urgent calls. In the world of interconnected project communities, it is easy to be accessible at all times. During your career as a Project Engineer you will often hear:

- "I never have enough time!"
- "I am constantly interrupted!"
- "I have to multitask continuously!"

Sounds familiar...Yes? I agree that as a busy Project Engineer, you will probably never have as much time as you wish, but armed with the tools described in this chapter, you will be able to maximise your effectiveness within your available time.

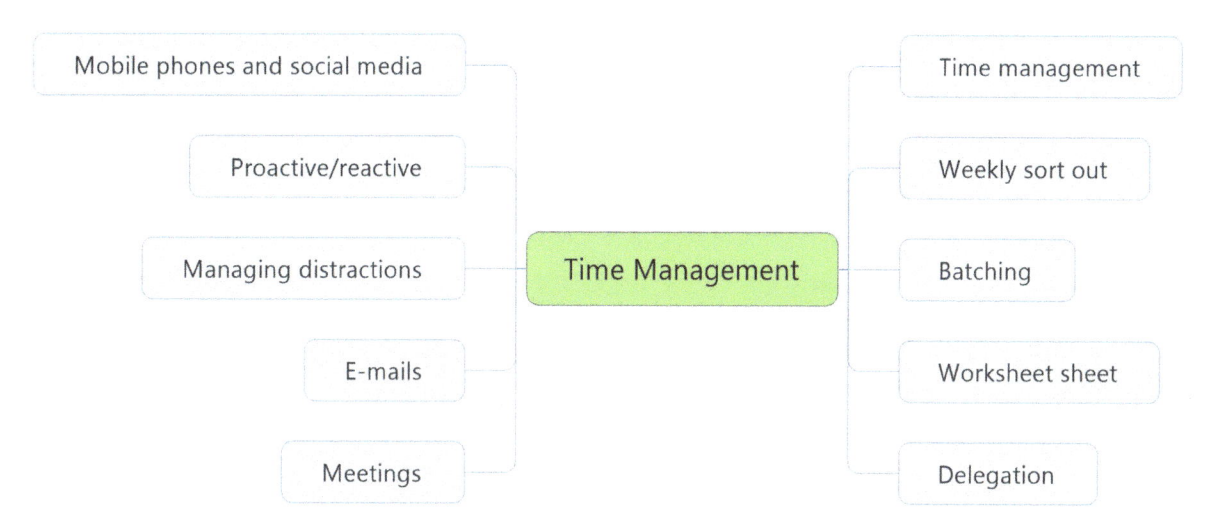

Figure 3.1 Time Management Mind Map.

Sure, we are busy, and we like to be seemingly indispensable, but stop and think: Are you truly being effective and efficient?

Managing the "Chaos of Inputs"

As Project Engineer, you will have continual multiple demands on your time and will be expected to filter and manage a "chaos of inputs":

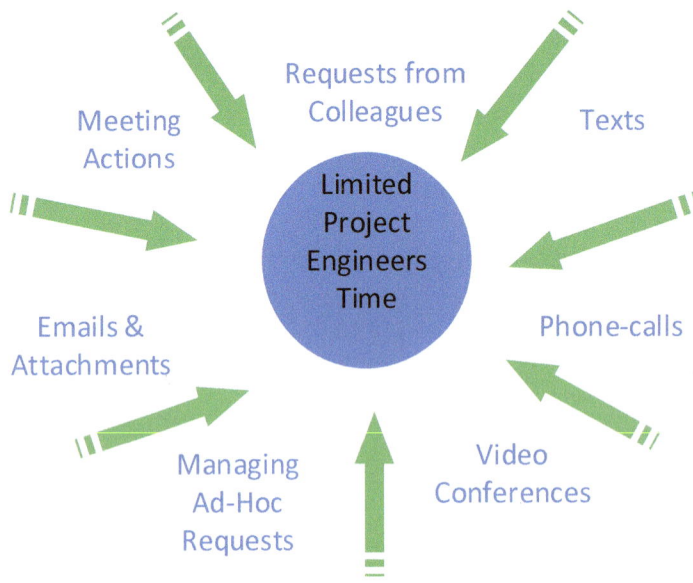

Figure 3.2 "Chaos of inputs".

One of the key skills to develop as a Project Engineer is effectively managing your time. It is not a case of working harder or longer but working smarter. If you don't, you will be at risk of failing to deliver and "burnout".

The inefficiencies of multitasking

Research (Gorlick, 2009) has shown that too much multitasking will actually slow down your cognitive processing. You will be unable to organise your thoughts or filter out unnecessary information. As a result, your efficiency plummets alongside the quality of your work. You are effectively spending your mental energy getting up to speed with an item, only to put it down before you have done any real "work" on it.

Multiplication of effort

As a Project Engineer you will be managing numerous resources. Hence, the better you manage your time, the more comprehensively you can manage and support your teams with a "cascade" improvement effect.

Time Management Methodology

The following section is intended to provide a methodology to allow you to incrementally improve your time management skills. As all the tools in this book, you will use the basic tool and shape it to your own specific requirements. The more you practice it, the better you will get. The following simple steps are suggested as a basis to develop your own system. Practice this religiously for six weeks and you will not go back to your old ways. I promise!

Weekly sort out and de-clutter

If your desk and e-mail files are cluttered, your brain will also be cluttered, and your project team will be less clear and focused. I have seen people proudly boasting that they have a thousand e-mails in their inboxes, and their desks are stacked high with paper. They advise that they are too busy with "real work" to organise themselves. In reality, they are just less effective and efficient.

Allocate regular time

Set aside some time every week, preferably the same time, to declutter and plan your time management. I set aside Friday mornings as there are fewer disturbances. This is one of the most important and effective things to do in your working week.

E-mail triage

Work through all your e-mails in your inbox. I call this "E-mail Triage". Decide what does not require any action or filing and delete them. Decide what needs done by you or a delegate, and add it to your action list. Finally decide what needs to be filed in your structured project filing system and add it in.

Hard copy triage

Repeat the e-mail triage exercise for your paper documents and working files. There will always many documents you can simply recycle.

Action list

You are now ready to generate your project action list(s). You will have a pile of actions, both electronically and physically:

- Identify actions that are urgent and important, and group them as priority one.
- Based on the client and management expectations and demands, prioritise these actions in the order they should be completed.
- Identify actions you must do yourself, and those that you can delegate to others.
- Go back to the priority one items and plan some time in your schedule to complete them.
- Group the delegated items into lists that you will be able to share with the wider members of your team.

What you have now just done is de-cluttered and removed the "noise". You will be feeling less stressed and have a clearer perspective of what is urgent and important, and when it will be achieved. Your wider team also benefits as you will be able to clearly advise actions and priorities.

Batching

Many people continuously juggle activities throughout the day and suffer from multitasking inefficiency. The trick to resolving this issue is batching. Only look at your e-mails twice daily and perform a mini-triage. Do not jump onto each e-mail as it comes into your inbox and feel the need to instantly reply.

If you are dealing with repetitive things like variations/change orders, do them all at the same time since they use the same thought process. Batch all your regular tasks to free up quality time for more value-adding activities.

CHAPTER 3 TIME MANAGEMENT

Weekly Worksheet

Project Engineering/Management can be tough. There will be short durations where 100% effort "sprints" are required. However, your project career is a "marathon," and you must take care of your prime asset: Yourself.

Most Project Engineers are time-limited and it can be very useful to plan out various aspects of your life and allocate the time each deserves. Set aside time for family, socialising, exercise, hobbies, etc. It is important not to neglect important areas of your life as this will also impact your work performance.

There is an excellent time management template in, "The 7 Habits of Highly Effective People: Powerful Lessons in Personal Change," by Steven R. Covey, upon which the following template is based. A download of this template is available at www.cranstoneng.com

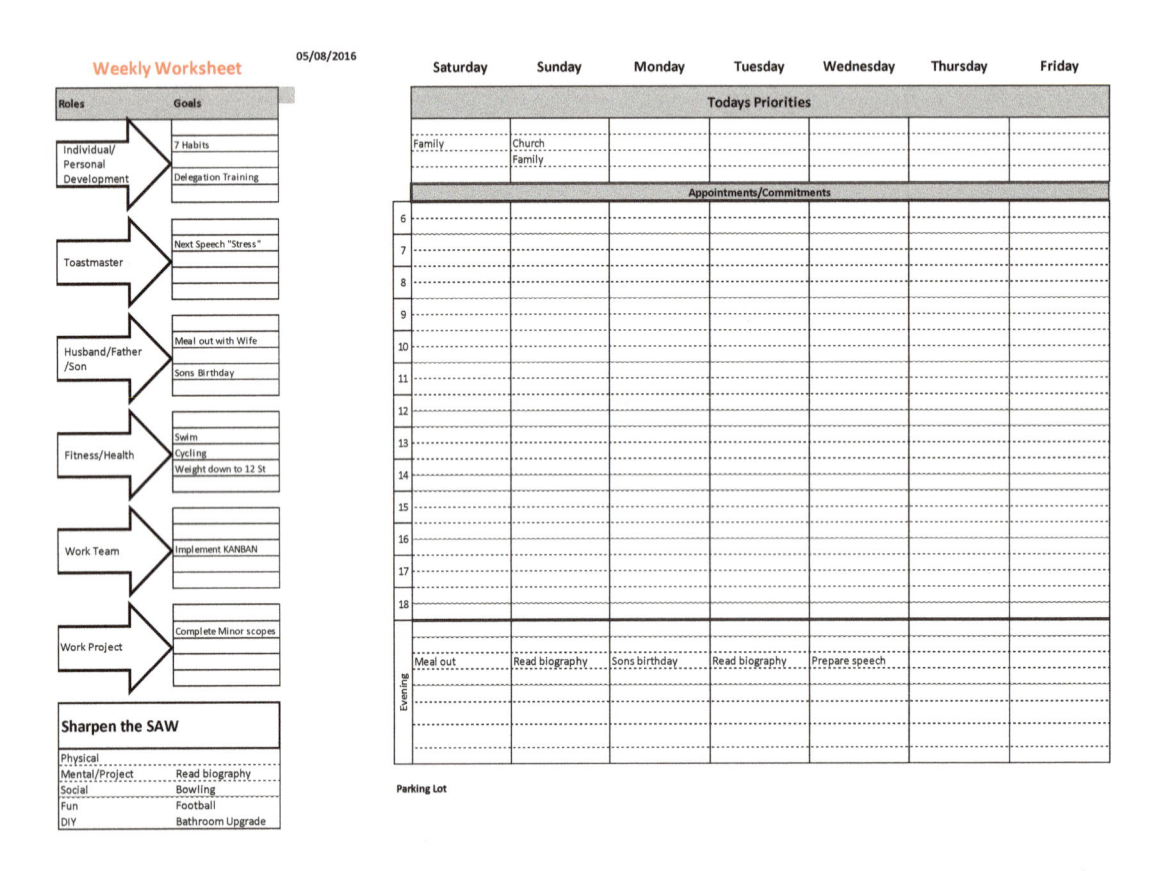

Figure 3.3 Weekly worksheet. Reproduced with permission from Franklin Covey:

Like de-cluttering, the preparation of the weekly time management worksheet can be practised and fine-tuned to best effect.

PETER F CRANSTON

Delegation

Delegation sounds easy but is one of the most challenging skills to develop. Correctly used, it is highly effective in multiplying your effectiveness and freeing up your time. More on this topic will be covered in chapter four.

Meetings

Being correctly prepared for meetings can be a very efficient use of time, and as a Project Engineer you will attend and chair many meetings. Imagine a series 1% improvement in all your projects meetings and the impact that it would have on overall project and organisation. Running effective meetings was covered in chapter two.

E-Mails

Incorrectly used, e-mails will waste an incredible amount of time. It is quite common to receive long e-mails and attachment trails copied to everybody, and in which you have to laboriously wade through to understand the real issues. Whenever possible, I much prefer in-person conversations or online meetings (such as with Skype) with visuals. We are designed for speaking to each other with instant feedback. Once information has been communicated, by all means follow-up with a short e-mail of agreements and actions for the record.

Managing Distractions

There will be many times, especially if working in an open plan office, where there are distractions from other colleagues talking, phones ringing, etc. When you have a task that requires lots of focus, consider working from home or finding a quiet room in the office. If it is not possible to do this, headphones with suitable music can help.

Proactive and Reactive

Proactive activities are planned in advance and are more efficient than reactive activities. Try to plan 70% of your weekly available time and keep the remaining 30% for unplanned reactive activities.

Do not try to plan 100% of your schedule as you will always fail due to the emerging reactive items. At the end of each week look back and see how many of the planned activities that you have completed. There will be times when there is simply more reactive activity, but planning ahead as much as you can will always be best.

Mobile Phones and Social Media

Mobile phones are a mixed blessing. You can be contacted at any time via voice, text, and many other systems. I keep my mobile phone at my desk unless I need to call people into meetings, or allow contact by reception, etc. Answering texts and taking calls in a meeting is just bad manners and raises the question of what value you are adding to the meeting if you are not giving it your 100%. If there is something really urgent, which is rare when you effectively manage your time, someone will come to the meeting room and call you out.

You can get Twitter, Facebook and many others feeds. There is an instant endorphin "fix" in reading that text or looking at Facebook, which is why we feel driven to check these platforms often. The downside is that this is wasting valuable company time. Try to ration your looking and responding to social media to once or twice per day in order to increase your productivity.

Time Management Summary

Practice the above time management principles and you will see continual improvement. More importantly you will feel less pressured and you will be more effective.

"In Projects you will normally be able to get more resources, and potentially budget but once a date has passed you can never get that time back. Therefore, spend time wisely."
- Ian McKnight

CHAPTER 4

Delegation

O kay, I have recently been made responsible for a significant Project. In previous jobs, I have done almost everything myself, but now I have a much bigger scope and I will have to delegate. Delegation sounds easy, but of all the tools in this book, it is the most challenging to master and do effectively.

Figure 4.1 Delegation Mind Map.

What is Delegation?

Delegation is the process of transferring the responsibility for a specific activity or task from one person to another and empowering that individual to accomplish a specific goal.

Project Engineers achieve their goals not by working harder and longer, but by delegating their work to others. When PEs delegate, they empower themselves and their team members. When the number of people you can help is limited, your success is limited, so when delegation is used effectively, everyone benefits.

Barriers to delegation

Sometimes PEs may feel uncomfortable about delegating for several reasons:

```
Barriers to Delegation ─┬─ Lack of Confidence
                        ├─ Control
                        ├─ Selfishness
                        ├─ Insecurity
                        └─ Reluctance
```

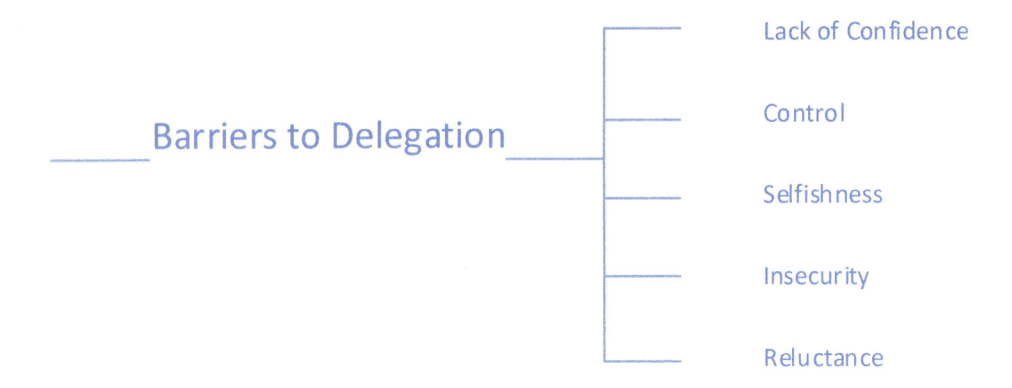

Figure 4.2 Barriers to delegation.

Lack of confidence

Some PEs simply do not believe team members have the skills and experience to do the necessary tasks. The team member may not initially have 100% of the abilities and skills to do the task, but with correct support and guidance, they will be able to complete the scope. There may be a learning curve, but once surpassed, they will be able to complete similar tasks with minimal support and supervision.

As the delegating PE, you will have to invest time and energy in supporting your team members as they develop the required skills. As you practice your delegating skills, you will be better able to judge what to delegate, to whom, and the level of support that they will require.

Control

Sometimes PEs are afraid of losing authority and control. If you try to micromanage and control everything, you will fail to develop the wider team and achieve the overall goals. Micromanagement is often resented and may cause frustration and conflict between team members. Conversely, if you do not have a good grip on the scope and who is doing what, you will equally fail.

I find when initially delegating to a new team member that daily briefings and feedback sessions with each individual team member works well. It is an efficient use of both your and their time. As confidence in each other grows, the duration and frequency of these meetings can be reduced.

Don't look for long, weighty status reports from your team. Encourage short, concise reporting with key points, issues, and required support.

Selfishness

Some PEs don't want to share credit. Fortunately, people like this are in the minority. If you are a PE and inclined to "hog the credit," think how much better your team will respond if you give credit where credit is due.
Team members who receive credit for work well done will respond by achieving even greater things, ultimately reflecting on your ability to lead a team.

Insecurity

Some PEs fear that a particular team member may do so well that he/she is a threat to the PE's job. There will be times when you have someone working for you who is "super competent" and completes requests quickly and with ease. They are clearly not going to stay at the level they are at.

Instead of seeing them as a threat, take it as a personal challenge to teach them all you can while they are part of your team. As they move up through the organisation they will remember who gave them important knowledge and skills, which is good for them, the organisation and yes, even you.

Reluctance

A few PEs are reluctant to ask others to take on additional responsibility, so they end up doing all the work themselves. You may be uncomfortable asking people to do things, which is not uncommon in new Project Engineers. You need to remember that some of your team members may have over 25 years' experience in being delegated to and expect it from the PE.

Remember that you cannot do it all yourself. As a Project Engineer, you will be expected to not just delegate, but delegate effectively. The following seven steps should be followed systematically until delegation becomes second nature.

Seven Steps to Delegation

Prioritise Actions → Match Needs to Availability/Ability → Assign Responsibility → Empower/Grant Authority → Establish Accountability → Follow Up → Recognition

Figure 4.3 Delegation process map.

1: Prioritise

Review the overall workload and identify what can be delegated. The input to your delegation process will have arisen from the long list of activities that you produced in your time management exercise. Identify those that you will or must complete yourself and prepare to delegate the rest.

2: Match needs to availability and ability

Match the requirements of each responsibility or activity with those who are available and can handle it. Consider individuals who are:

- Knowledgeable. Make sure the team member has the knowledge needed to do the task.

- Motivated. The team member may be knowledgeable, but if he/she does not believe in the Project Engineer's goals or just isn't interested, it is unlikely that person will make the effort needed to accomplish the task at hand. When possible try to convince them why this activity should be done.

- Has Capacity. If the team member has other responsibilities not related to the Project Engineers goals, he/she may not have the time to devote to the assigned task or may not be able to give it the attention it needs. To give that team member capacity you may need to negotiate priorities with management or defer the activity until the team member has the capacity in his/her schedule.

3: Assign responsibility

You will need to clearly explain the tasks for which the team member is responsible for as well as the expected results.

Fully Explain Task → Provide all Required Information → Advise Supporting Parties → Identify Training Needs → Identify Resources

Figure 4.4 Assigning delegation responsibility.

Follow these steps:

- Fully describe the task and expected results.
- Give all the information necessary to get the task started, or tell the team member where it can be obtained.
- State who else will be involved and explain their roles.
- Identify training needs. If the team member needs special training or help to accomplish the task, provide it.
- Identify resources. Make sure the team member knows the materials, information, and budget available.

The above is the investment in time and energy you will make to enable the team member to deliver. This should always be followed up in writing as a record of the delegation request.

4: Empower/grant authority

Give the team member the authority to gather resources and make the necessary decisions to achieve the desired results. This may be advised at a team meeting or via an e-mail instruction. It is important to specify the amount of authority being delegated. For example, if the team member must get your approval for expenditures over $1000, be sure he/she knows this.

5: Establish accountability

This is the important bit. Many novices in delegation skills simply issue an instruction and are disappointed and frustrated when it is either not done or not done as requested. You should hold team members accountable for completing the tasks they agreed to complete. Agree on performance standards and timetables for completion. Prepare a report timetable. State the amount, frequency and types of progress reports desired. Take care not to ask for too much detail otherwise too much effort will be put into reporting rather than the actual activity itself.

6: Follow-up

Follow-up is vital in the delegation process. As part of accountability, PEs must ensure that team members know to share both successes and problems. PEs must offer feedback and help when necessary. Results will not meet expectations unless the PE pays close attention and monitors progress.

When delegating someone you have not worked with frequently before, communicate daily to align understanding. You can reduce this communication as mutual trust and respect develops.

7: Recognition

The Project Engineer should recognise team members for their achievements when they provide the desired results. Recognition encourages the specific member as well as the entire team to aim higher as well.

Recognition should be appropriate to both the culture and the persons involved. While "high fives" and public meeting recognition might work well in some cultures, others might find this embarrassing and shun the publicity. Recognition may be in the form of an article in the project or company newsletter. Some projects have budget that allows meals or other nominal sum awards for delivering results.

An e-mail to the person's line manager advising on the good performance can also be effective and taken into account at annual staff reviews. One of the most powerful forms of recognition is a simple, genuine heartfelt "thank you" to the team member for their efforts and performance in delivering.

Recognition should be given only where deserved and, correctly used, will motivate members to even higher performance

Types of Delegation

Delegation can be horizontal or vertical.

Types of Delegation
- Upwards
- Peer
- Downwards

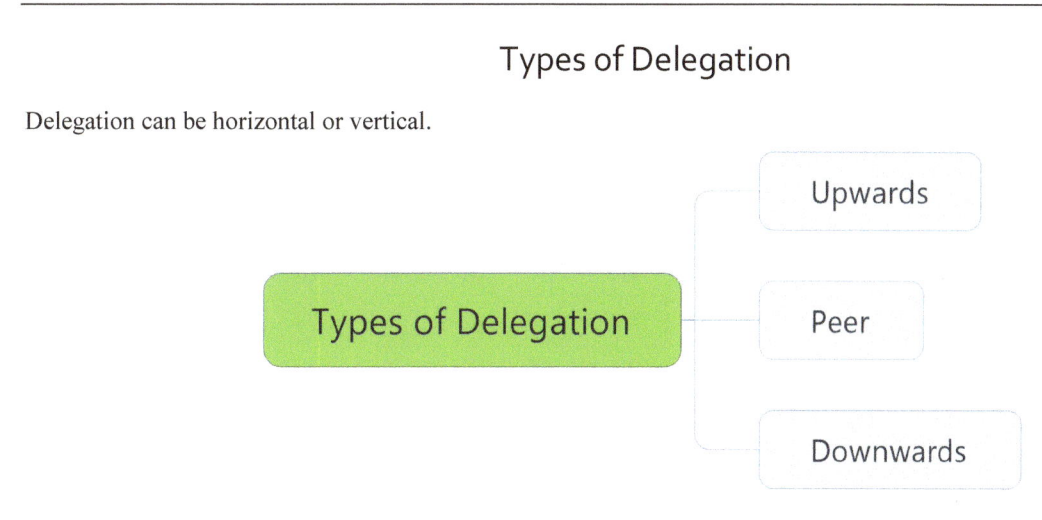

Figure 4.5 Types of delegation.

Downwards delegation

Downwards delegation is the most common form of delegation a Project Engineer will use to provide instructions to a team member who will in turn report on status and completion.

Peer or horizontal delegation

There will be times when you will need your peers to assist you in completing elements of a project, and it can be initiated by either party.

Upwards delegation

Upwards delegation is used sparingly when your boss must complete an activity. For this it is important to have a good relationship with your boss.

Evaluating your Workload Percentage Capacities

Aligned with the skill of delegation is evaluating you and your team's workloads. Effective team members are good at what they do and can get loaded up with more scope and activities that can reasonably be completed within the required schedule, causing stress.

An effective technique to use for yourself and your team members is to write down the tasks and then honestly estimate the percentage (%) of the working week that will be required to complete the task. Then, sum the percentages of all the tasks. If the sum is above 100%, you need to offload some scope or delegate the task to another team member.

Percentage Capacity Checker

List of Tasks	% Workload
Weekly Report	15
Project A Engineering	20
Project B Study	40
Project C Construction Support	30
Planning Meeting Prep' and Chair	10
Project Change Control Activities	10
Offshore Team Briefings	10
Daily Site Conference Calls	10
Required Capacity	145
Available Capacity	100

Figure 4.6 Percentage capacity checker.

If your workload is significantly above 100%, it will require upward delegation from your boss to resolve. A wise boss will appreciate you bringing this to their notice as overall delivery of the entire project/program is his/her responsibility.

Benefits of Effective Delegation

Good delegation requires thought and careful planning, and has many benefits. When done properly, it can lead to:

A more involved and empowered Workforce

Good people like challenges and respond well to being given responsibility, allowing them to enjoy their work more.

Increased productivity and quality

Overall your project will become more productive with a likely rise in the quality of work.

Reduced costs

Increased productivity means more work is completed in less time, and increased quality means less re-work, both of which will reduce costs and help increase profits.

More innovation

When someone is given a delegated responsibility and the resources to complete the task, it is wise to allow him/her some flexibility within agreed limits. By delivering on the scope using his/her own approaches, the team member is more likely to find better and innovative ways to complete the task.

Greater commitment from team members

Team members are more likely to be committed to a goal when they are involved and allowed to contribute to its success. Once people are actively involved in delegated projects and activities, including problem-solving, they develop a sense of ownership.

Effective delegation benefits the Project Engineer too. By delegating tasks to team members, PEs can use their time to accomplish more complicated, difficult, or important tasks and goals. This can lead to a more creative and successful organisation as a whole.

Delegation in Practice

Many years ago, I was allocated a project to manage a major offshore shutdown on a North Sea Platform. The scope was to up-man to 450 using a dedicated Flotel for a six week period. During this period it was planned to:

- Carry out major process modifications to convert the platform from 2 to 3 phase separation and to isolate the concrete oil storage cells.
- Complete drilling facilities upgrades.
- Carry out subsea field tieback maintenance.
- Implement a maintenance program.
- Prepare and manage mobilisation of the Flotel.

CHAPTER 4 DELEGATION

The initial challenge was to decide how to manage and coordinate these different groups. It was clear that the meeting minutes and detailed action logs would be too unwieldy. The selected solution was to allocate a single focal point for each of the groups, as shown below.

Figure 4.7 Flotel campaign management focal points.

Regular interface meetings were then set up with each focal point and a customised data base was used for all the actions, allowing for easy filtration of actions against certain criteria and effectively delegating action lists for the focal points. Each focal point had clarity on its actions and action history.

The focal points then sub-delegated actions as needed. Due, in part, to this delegation methodology the shutdown was completed safely and within schedule. I advocate the use of customised databases for any project where there will be significant delegation and complexity.

Delegation Summary

You as a Project Engineer will be evaluated not only on your individual performance, but also on how well your teams perform. A PE who takes the time to develop delegation skills is more likely to have a successful team.

"The best executive is the one who has sense enough to pick good men to do what he wants done, and self-restraint enough to keep from meddling with them while they do it."
- Theodore Roosevelt.

CHAPTER 5

Conflict Management

The objective of this chapter is to explain the sources of conflict in a project environment and provide you with techniques to manage them.

Figure 5.1 Conflict Mind Map.

Definition

There are numerous definitions of conflict:

- A serious disagreement or argument, typically a protracted one.
- An active disagreement between people with opposing opinions or principles.
- Friction or opposition resulting from incompatibilities, or actual or perceived differences.

In a project environment, conflicts can be between individuals or groups of people.

CHAPTER 5 CONFLICT MANAGEMENT

Why Conflict Should be Managed

Conflict is inevitable in projects, but if managed correctly can improve the overall team performance. Some people shy away from conflict in a project environment, especially if they are not naturally forceful. If conflict is not addressed early and properly, the project will suffer.

Inevitable on projects

Conflict will happen in situations involving more than one person. Given a project environment with numerous people, different backgrounds and agendas, conflict is a guarantee. It is an important soft skill to manage conflict.

Good and Bad Conflict

Early management theories viewed all conflict as negative, however the modern perspective [Ref1] is that correctly-managed conflict in a project can be good.

Good conflict

Good Conflict has the following characteristics:

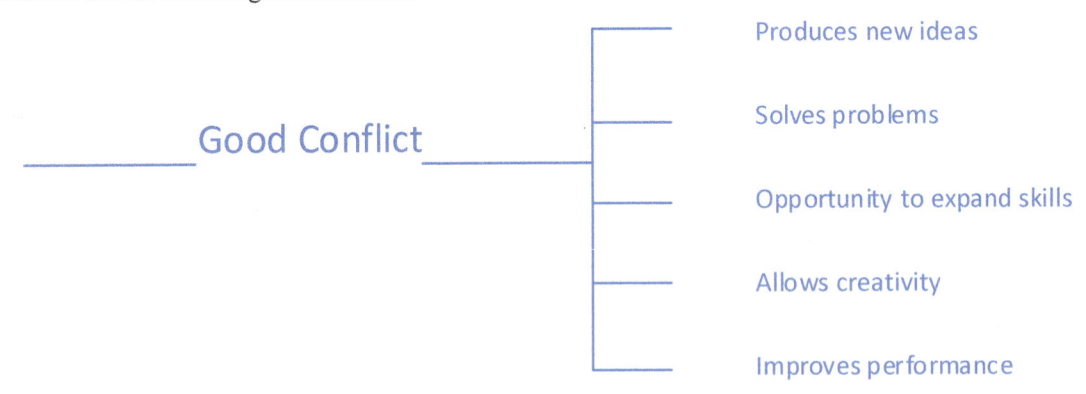

Good Conflict
- Produces new ideas
- Solves problems
- Opportunity to expand skills
- Allows creativity
- Improves performance

Figure 5.2 Good conflict characteristics.

Bad conflict

On the other hand bad conflict can result in the following:

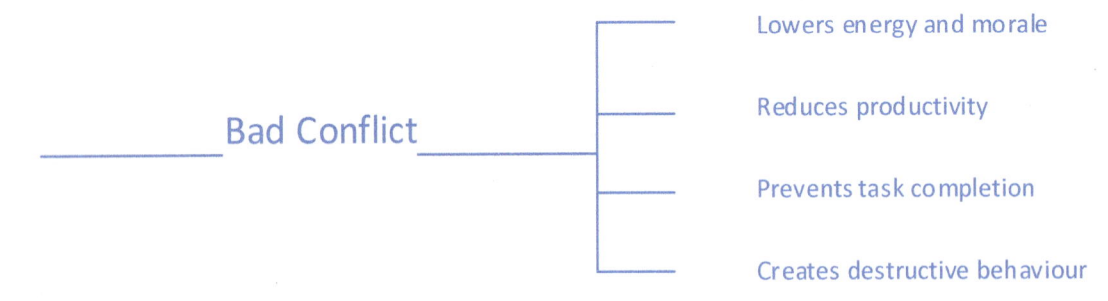

Bad Conflict
- Lowers energy and morale
- Reduces productivity
- Prevents task completion
- Creates destructive behaviour

Figure 5.3 Bad conflict characteristics.

Reasons for Conflict

There are seven generally accepted major sources of project conflict[Ref2]

```
                                              ┌── Schedules
                    Personalities ──┐          │
                                    │          ├── Project priorities
                         Costs ─────┤          │
                                    ├── Reasons for Conflict
                                    │          ├── Resources
        Administrative procedures ──┘          │
                                              └── Technical options
```

Figure 5.4 Sources of project conflict.

It is interesting to note that personality conflicts are the least common sources of conflicts in projects.

Symptoms of Conflict

The most common symptoms of conflicts in a project are listed below:

- Absenteeism.
- Complaints.
- Gossip.
- Withholding information that should be shared.
- Hostility.
- Not attending meetings.
- Not completing agreed deliverables.
- Not responding to requests.
- Verbal aggression and abuse.

Characteristics of Low Conflict/High Performance Projects

- Effective communication by all team members.
- Teamwork to achieve mutual goals.
- Sharing of achievements.
- Meeting desired goals.
- Working together.
- Resolving conflict.

- Mutual respect between all team members.
- Understanding project roles and responsibilities.

Suggested task: If you are in an active project role, review the lists above and aim to identify both the good performance and conflict indicators. You may find that certain characteristics are more or less exhibited by particular individuals or groups.

Cultural Differences

It is not uncommon for projects to have teams comprised of many different cultural backgrounds and are either co-located or distributed.

Cultural conflicts can arise because of the differences in values and norms of people from different cultures. A person behaves according to what is appropriate to his/her culture. Another person with a different worldview might interpret his/her behaviour from a different standpoint, and what is acceptable in one culture may be totally unacceptable in another. These situations create misunderstandings that can lead to conflicts.

Cultural conflict can arise from generational differences, educational levels, and ethnicity, and cause strong emotions even when insignificant issues. The two keys to minimising cultural conflict are understanding and respect.

Understanding

Take time to understand the other person's values and customs. I have found that talking to people in the workplace about their cultures is both interesting and educational on a personal level, and in return I have been able to share some of my own culture. Both the project itself and personal relationships with team members benefit.

Respect

Respect other culture's values, even if you do not agree with them.

Virtual Team Considerations

By definition, virtual teams are located in different locations and potentially even different time zones. While this might reduce costs and increase resource flexibility, the lack of face-to-face interactions can generate conflict.

The following actions are recommended to build trust and personal interaction in virtual teams:

Initial face-to-face meetings

At the start of the project, try to get key team members together in the same location. If this is not possible, use video conferencing. When you have met someone in person, you are more likely to communicate more effectively later via phone calls, video conferencing etc.

Set up a social network

Set up a social network such as Yammer[Ref3] to allow team members to interact socially while working on the project. I have used this successfully a number of times.

Facilitate team member partnerships

Facilitate the partnership of team members working in different locations to encourage collaboration and interaction.

Exclusive virtual team member meetings

Schedule exclusive time, in-person or virtually, with each team member to seek feedback and allow them to air any concerns.

Resolution Strategies

Conflicts will occur in projects so it is useful to have some tools to resolve them.

Solve conflict early

Whilst most people tend to delay resolution of conflict, an early resolution is best for project success. Unresolved conflicts can fester, multiply and cause significant damage.

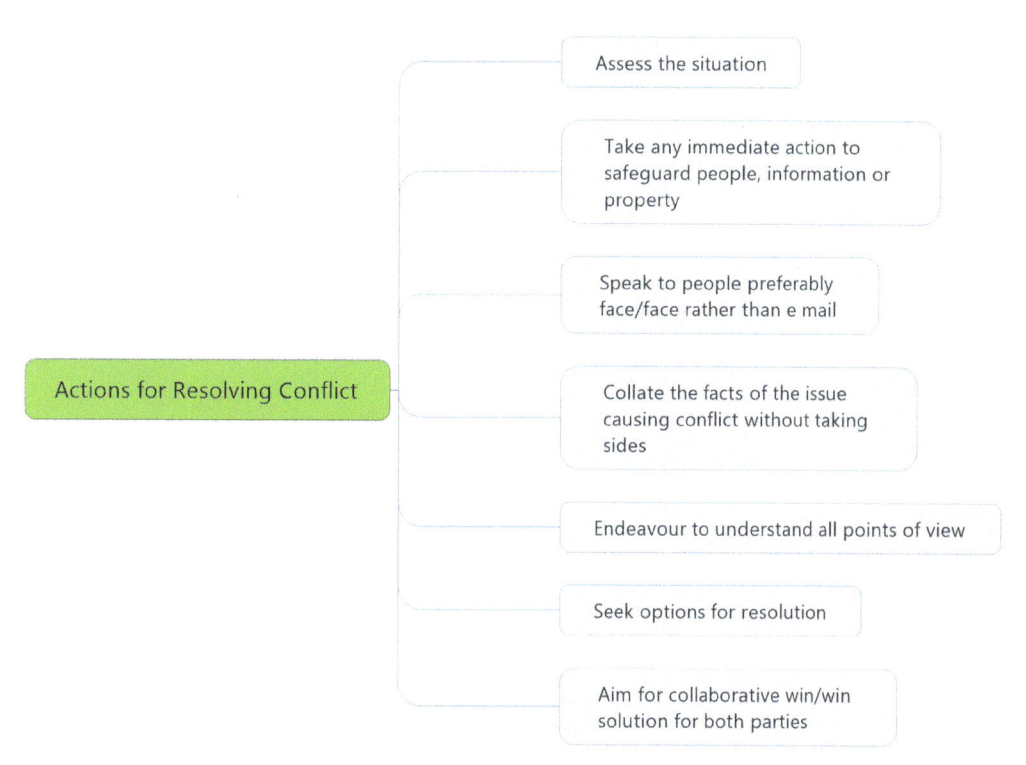

Actions for Resolving Conflict:
- Assess the situation
- Take any immediate action to safeguard people, information or property
- Speak to people preferably face/face rather than e mail
- Collate the facts of the issue causing conflict without taking sides
- Endeavour to understand all points of view
- Seek options for resolution
- Aim for collaborative win/win solution for both parties

Figure 5.5 Steps for resolving conflict.

Interpersonal conflict

When two people are in conflict, encourage them to discuss and resolve it themselves. If they are unable to do so, act as an arbiter between the two parties, and in all cases, maintain professionalism and do not take sides.

Intergroup conflict

Intergroup conflict can be reduced by seeking to identify the key influencers in each group and seeking their support in airing and resolving issues. Intergroup conflict can also be reduced by setting expectations of behaviour regarding team member support and respect. Removal or reassignment out of the project may be required for any parties not complying with required behaviour standards.

Clear priorities

One of the greatest sources of conflicts is priorities. To reduce this, provide clear priorities to all your team members ensuring that the priority clashes are minimised. In Matrix environments, expect to spend a significant amount of time agreeing and managing priorities with service providers, Project Managers, and clients.

Clear communication

Aligned with clear priorities is clear communication. If everyone is "on the same page" regarding scope, schedule, issues, and so on, there will be fewer "clashes of understanding".

Team building

Team building sessions are great for airing issues and discussing topics in an open environment. Coupling meals with team building activities can encourage people to talk about wider work and social items. Team building sessions are normally held at the start of each new phase of a project, however, if the project has a conflict problem, a carefully prepared and facilitated team building session can address and reduce the conflict.

Selective change-out

Project teams are formed from a wide cross-section of people, all with different personalities, likes, and dislikes. There will be times when a person has a sufficiently negative conflict effect on the project that replacement will be required for the benefit of the project. It may have simply been a mismatch and this same person may not be a source of conflict on another project. It goes without saying that all selective change-outs should be managed sensitively to avoid demotivation.

Example of Conflict Resolution

The new engineering manager

Some years ago, I was the Project Engineer of an offshore platform accommodation project. The project was running smoothly with regular weekly project meetings, a detailed agenda, and a robust structure to the meeting with various team members responsible for agenda items.

A new engineering manager was appointed to the project and was invited to the regular weekly project progress meeting. Within minutes of the start of the meeting, the new manager aggressively and verbally attacked and offended a number of the team members, effectively destroying the smooth running of the meeting. As meeting chairman, I stated that the meeting was now unproductive, closed and would be rescheduled.

I then immediately sought out the Project Manager and briefed him on the conflict in the meeting. After a three-way discussion, heated at times, between the new Engineering Manager, Project Manager and I, we agreed that the Engineering Manager would not attend the projects' progress meetings, and that he would instead chair a separate engineering meeting focusing on engineering issue recovery actions. I also took time every morning to informally discuss status and issues with the new Engineering Manager in his office.

Despite the difficult relationship with the engineering manager, the above agreements allowed us both to carry out our required duties. There will always be some difficult people on a project and the challenge is to find what works for you and the project.

Conflict Summary

Conflict is inevitable in a project and will have to be dealt with even though most of us do not find it enjoyable. As leaders on projects, Project Engineers will often witness or be involved in conflict.
One of the key measures for how successful a Project Engineer is how effective he/she is in defusing and managing conflict in the team.

"Whenever you're in conflict with someone, there is one factor that can make the difference between damaging your relationship and deepening it. That factor is attitude."
- William James

Chapter 6

Motivation

Before you read this chapter I would like you to take some time to jot down your personal views on the following three questions.

- Why do we need to motivate?
- Are you a motivator?
- What do we need to do to motivate our teams?

Figure 6.1 Motivation Mind Map.

Getting your project team properly motivated has many benefits, such as being more productive, which will positively impact the bottom line, being more likely to innovate in both technical and work process areas, and creating an overall more enjoyable experience for everybody, promoting a positive "buzz" that lifts the whole team.

Project Structure

Some team structures are harder to motivate than others. Consider a dedicated project team under one Project Manager in which the entire team is aligned to the project delivery. Then, consider a matrix team servicing multiple projects and Project Managers. It will be harder to motivate the latter team when they have many customers to manage, probably all with urgent and important deliverables.

This chapter will review motivation theory, motivators, and de-motivators. At the end of this chapter, I suggest you complete a self check using the following list of motivators and demotivators and reflect on both the positive and negative motivators evident in your project(s).

Motivation Theory

There are a number of motivation theories, with the most well-known being:

- McGregor's Theory of X and Y.
- Maslow's Hierarchy of Needs.

McGregor's theory of X &Y[Ref4]

Theory X	Theory Y
People	**People**
• Need close supervision	• Want independence in work
• Will avoid work where possible	• Seek responsibility
• Will avoid responsibility	• Are motivated by self-fulfilment
• Only desire money	• Naturally want to work
• Must be pushed to perform	• Will drive themselves

Figure 6.2 McGregor's theory of X and Y.

Theory X is based on pessimistic assumptions regarding the typical worker. This management style supposes that the typical worker has little to no ambition, shies away from work or responsibilities, and needs to be watched all the time. Whilst this may be partially true in a low skilled production line environment, it is not true for the vast majority of people working in the Oil and Gas industry in a project environment.

Theory Y, on the other hand, assumes that people are motivated to do good work and enjoy producing it. They take much more responsibility for their work and require minimum supervision. Project Engineers and Project Managers who follow this theory generally have much higher success and a more contented team than those who follow the pessimistic Theory X.

My personal approach is to initially treat all the team according to Theory Y. If they do not live up to these principles, I then coach improvements in behaviours and attitudes. If, and only if, this does not work do I consider selective replacement.

Maslow's hierarchy of needs[Ref5]

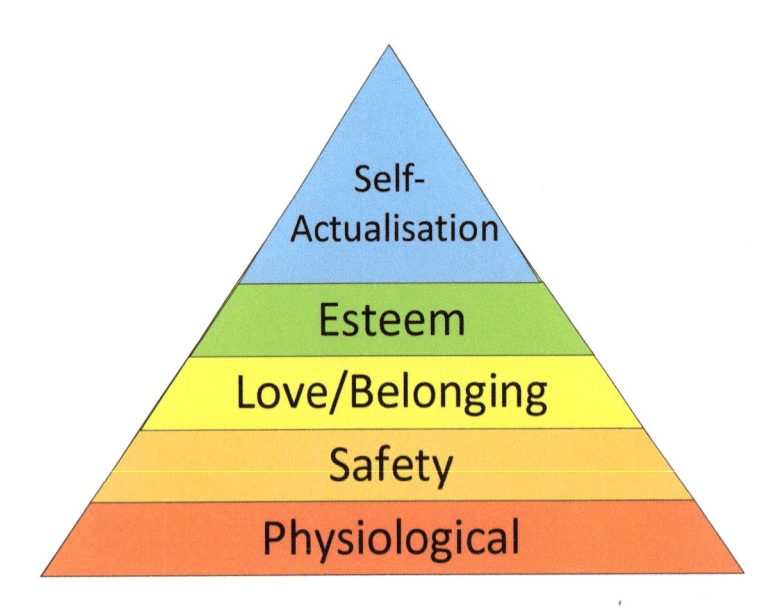

Figure 6.3 Maslow's hierarchy of needs.

Maslow's Hierarchy of Needs is a theory in psychology proposed by Abraham Maslow in 1943. Simply put, it proposes people are motivated to achieve certain needs and that some needs take precedence over others. Our most basic need is for physiological survival and this will be the first thing that motivates our behaviour. Once that level is fulfilled, the next level up is what motivates us, and so on.

Armed with the aforementioned theories, we can now look at motivators that can be used to inspire your teams. You may not always be able to directly apply these motivators due to various constraints, but when possible, they should be considered.

Motivators

Money

Like it or not, many of us are motivated by money as we have mouths to feed and bills to pay. However, money alone does not provide lasting motivation.

Recognition

People like to be recognised for good efforts, and thanking them genuinely in public can give them a boost. Sending an e-mail of thanks and copying their managers gives recognition and the possibility of a pay increase at the next review. A simple honest "Thank You" costs nothing.

Responsibility and promotion

Giving a keen team member more responsibility allows him/her to develop professionally and brings along the potential for future promotion within the organisation. Regarding future promotions, the team member should be advised of the realistic expected time period to promotion to keep the motivation strong.

Alignment of role with development aims

Some of the most effective motivators I have seen demonstrated is when team members are given roles and responsibilities that align with their personal development aims, such as construction site experience, tender preparation, exposure to senior management, and so on.

Focused training and coaching

Giving your team the skills to do their jobs more effectively is really important, and time spent coaching will pay dividends. One-on-one coaching in identified areas where support and guidance can be provided is priceless. If you are offered either formal or informal coaching, take advantage of it.

Honesty

Be honest with your teams. They will come to respect you as someone "on the level" who can be trusted. Do not make promises that you cannot keep.

Fairness

As a Project Engineer, you need to be fair in your dealings with everyone, be they clients, project team members, or suppliers and subcontractors. Being fair will build up your reputation as a person of integrity.

Favouritism

Avoid having favourites and giving them preferential work scopes or working conditions based on favouritism alone, as it will be noticed and cause dissent in the team.

Rewarding desirable behaviours

Reward those who perform well. Each project will have its own acceptable rewards, which may include wider recognition in company newsletters, an appropriate gift, or a monetary award.

Punishing undesirable behaviours

Equally when someone does not perform, let him/her know what the expected performance or behaviour is as well as the consequences if the behaviour is not corrected. You should be sensitive to reasons for undesirable behaviours, such as personal relationships or health issues and provide support when possible.

Demotivation

A demotivated team will fail to deliver either on time or on schedule. A project can be challenging even with a fully motivated team. The following are areas to watch out for in your projects:

Unresolved conflict

Conflict will happen on projects and it is only when unresolved that demotivation can creep in. Seek early conflict resolution using conflict management techniques (See Chapter Five).

Loss of respect

It can take months to build up mutual respect with your team members, but this respect can be destroyed in just a few seconds with inappropriate, harsh words or treatment. The offended party may "switch off" and lose their motivation, potentially affecting performance and quality of work. It will take huge effort to rebuild the trust and motivation.

Project saboteurs

On any project. a "project saboteur" can wreak havoc. This can usually happen in larger projects where there is initially some anonymity. The saboteur is typically an intelligent team member who may have a personal agenda with some of the project team members, or simply has a bad attitude. He/she will respond negatively to everything and, as a result, will negatively influence the wider team. It is not usually clear at the start of a project who the saboteur is, but when recognised, he/she should be transferred off the project since he/she is costing the project and the organisation money.

Stealing credit

Always give credit to the people who have delivered and never take the credit for others' work. As a Project Engineer, you are only as good as your team, so spend time giving thanks and credit when appropriate.

An Example of Demotivation

Whilst working on a module project, we had an Engineering Manager who was cheerful, friendly, and diligent. Shortly after I arrived, a new Project Manager was appointed who was clearly out to make name for himself. He was aggressive in the extreme; a real "table thumper" who berated the engineering manager publicly in client meetings. The previously energised and helpful engineering manager went on the decline and lost all motivation, to the extent of leaving the organisation and the industry after 9 months of such horrendous treatment.

An Example of Motivation

One project required prototyping a piece of equipment with full-scale onshore trials for machining hatches in a Zone 1 offshore hazardous area. I was supported by a junior Project Engineer whose previous experience was mainly project change control. He was very keen on engineering, but as a project management graduate was unable to obtain any practical engineering experience.

Since I knew his interests, I involved him fully in the technical aspects of the project and the subsequent onshore workshop trials as part of his development. He really enjoyed the experience and exposure to prototype engineering and the payback to the project was his tireless work in support of the project. A "win-win" for all parties.

Motivation Summary

What I have learned over my career regarding motivation are two things:

When possible, provide at least one element of what team members would like in the working environment. This may be an opportunity to go to an offshore site. It may be some coaching or training. It may be a recommendation for a position on a new project. It may simply be reduced working hours to allow for family commitments. The trick is to identify what motivates people, what is possible to accomplish it, and fairly apply it.

Secondly, maintain an upbeat positive attitude with the entire team as positivity engenders positivity in others. Positive people always do better.

"Our greatest weakness lies in giving up. The most certain way to succeed is always to try just one more time" - Thomas A. Edison

Presentation Skills

Some people love giving presentations while others dislike doing them. Whatever your current presentation skills and confidence, presentations are one of your most important opportunities to impress and communicate clearly to clients, superiors and colleagues. As a Project Engineer you will need to frequently communicate to groups of varying sizes. This chapter will provide a structure on how to prepare for, practice and deliver winning presentations.

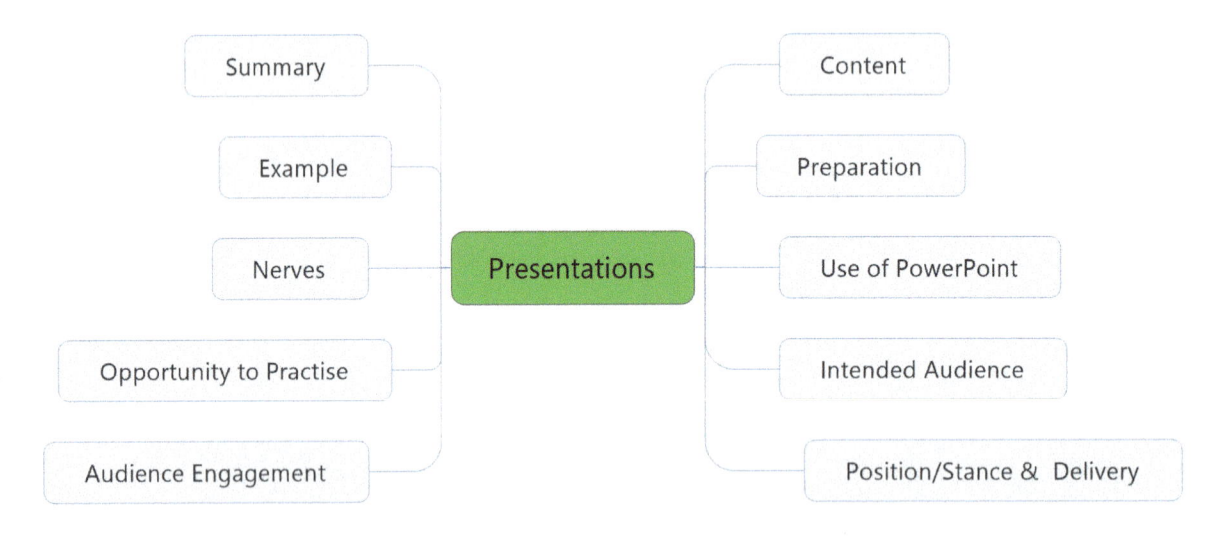

Figure 7.1 Presentations Mind Map.

Content

You need to know your subject because nothing comes across more clearly than lack of knowledge. Chances are you have been asked to deliver a presentation because you are knowledgeable and possibly the expert on the topic being presented. However thorough preparation and checking facts and figures makes all the difference.

Mindmap approach

I prepare a mindmap for every presentation. This chapter started out as a hand-drawn mindmap. Dump all your thoughts as you think of them and then set them down in a preferred or required order. When you are more comfortable with mind mapping, you can very quickly put your presentation structure together. Many software mindmap programs are available online, both free and for a fee.

A well-structured presentation organises your thoughts

A structured approach allows you to clearly communicate the subject matter and ensure you cover all items within the allocated time.

Preparation

The larger the audience, the higher the level of management, or the more important to your career, the more preparation is required to set up a presentation. I have run presentation courses and it is always clear who has prepared and who has not.

Dry run

Once drafted, book a meeting room and do some "dry runs" by yourself or with a trusted friend or colleague. Time it so you can adjust the length as required to fit your presentation slot.

Arrive early to the venue/room

Arrive to the presentation room early. If using information technology (IT), arrive at least 15 minutes early to ensure that it all works. Nothing piles on pressure than turning up last and trying to get IT, laptops, projectors and so on working. This also allows you to get the feel of the room and relax a bit. There will be more on managing nerves later.

Check IT

Nowadays, most work presentations will use PowerPoint although it is not always required. Load up your presentation and ensure it displays correctly on the screen. Also check the operation of laser pointers and mouse.

Notes

Prepare notes to ensure you cover everything, while also allowing you to get organised. These can be printed out or you can use the PowerPoint notes facility. If you know your subject, you will only need a light reference as you go through the presentation.

Backups

Equipment failures happen. Have your backup presentation on a colleague's laptop, use a memory stick, and, just in case, have a hard copy print out.

Without IT

Imagine you turn up and the IT has totally failed. A good presenter will have pre-printed his/her slides and note pack and, depending on content, will be able to deliver a reduced presentation without visual props.

CHAPTER 7 PRESENTATION SKILLS

Use of PowerPoint

We have all experienced "Death by PowerPoint". PowerPoint is only a tool to share information. The key impact of a presentation is the presenter him/herself. The purpose of PowerPoint is simply to provide supporting graphics and summary text.

Less is more

A text-busy slide is hard to read and detracts the audience from the speaker. With busy text, the presenter tends just to read the slides aloud.

Speaker Reads Slides

- A speaker may put his/her entire presentation on his/her slides. They turn their back to the audience and read the slides aloud. They may feel that this approach guarantees all the information will get to the audience.
- This is probably the most annoying way to give a presentation. Audience members feel insulted as they already know how to read.. They wonder why the presenter does not simply hand out a copy of the slides
- The visual presentation dominates the presenter. The presenter is not adding any value to the slides

"Note: This slide is way too busy!"

Figure 7.2 Example of bad practice.

The aim of the presentation is to communicate information, not swamp the audience with data. In presentations, light, clear, and concise slides with a narrative communicates more information that the audience will find easier to retain.

Good PowerPoint practice

Try to shorten bullet points:

- No more than 5 lines per slide
- Clear diagrams
- Zooming-in for details

THE PROJECT ENGINEER'S TOOLKIT•67

> **Good Presentations**
>
> - Short bullet points
> - 5 lines max per slide
> - Clear diagrams
> - Avoid "toys"
> - Less is more!
>
> Copyright © Cranston Engineering Ltd 2018

Figure 7.3 Good presentation practices

Avoid toys

I have seen people spending time adding animations and sounds to their presentations.
Ask yourself "What does this add to the information I am communicating?" In most cases, it does not add anything and instead is an annoying distraction. It can detract from the content and message you are delivering. I have, however, seen animations used effectively to show a sequence on a schematic. Transitions, animations, and sounds should only utilised when used sparingly or essential in adding value.

Back-up slides

You will have a limited time to present and will have timed your presentation to fit the available time. However, have back-up slides to illustrate or explain any anticipated questions or clarifications.

Intended Audience

Carefully prepare your presentation for the intended audience. Are they technical? Are they commercial? Is it the site workforce?

Too detailed?

Too many details turn people off and may cause you to overrun your allocated time.

Not detailed enough?

Too little detail will not properly inform or convince the audience and result in many questions or a view that you lack the appropriate knowledge. Similarly, consider if the content is too complex or too simple and adjust as needed. This is where performing a dry run and obtaining feedback from a trusted colleague can be very valuable.

Position/Stance & Delivery

As a Project Engineer, you are responsible for the project or element of the project that your clients and management have given you, demonstrating their confidence in your ability to take on significant responsibility.

Some simple tips

- Face the audience when possible.
- Stand if possible.
- Pause before starting and take a deep breath.
- Avoid placing your hands in your pockets, or fiddling with pens, glasses etc.
- Do not apologise.

Good practice

- Simple and concise delivery.
- Clear and paced delivery.

Bad practice

- Turning your back to the audience to read the slides.
- Speaking too quickly.
- Speaking too quietly.
- Mumbling.

Audience Engagement

Introduction of your topic

Tell the audience the objective of the presentation and share the presentation structure.

Eye contact

Maintain eye contact with the entire audience, working around the room. Different cultures vary in what is acceptable regarding eye contact, so be conscious of not making anyone uncomfortable.

Questions

Ask open questions to encourage interaction with your audience.

Avoid directly reading

Unless you need to quote specific text, such as an excerpt from an engineering standard, use your own words and paraphrase as needed. This will come across as more natural and less stilted.

Re-state what you said

Summarise and recap, recording any key feedback issues or actions. A flipchart and an assistant, if available, are useful.

Opportunity to Practice

Depending on the audience and importance of the presentation, one or more dry runs can be carried out, allowing for feedback from a trusted colleague on content modifications to establish overall timings. A good presenter finishes on-time.

Performance Nerves

Negative adrenaline

One of the most common reasons people do not wish to deliver presentations is not lack of knowledge but rather performance nerves. At a funeral, for example, many people would rather be in the coffin rather than standing up and giving the eulogy. Presentation nerves are caused mainly by adrenaline for the "fight or flight" response. This was useful in the Stone Age, but not so much in presentations. Here are a few tips to reduce and manage that adrenaline:

Room awareness

Arrive to the room early and stand at the screen or podium and visualise a successful presentation.
Practice relaxation and breathing exercises, using the internet to find a multitude of techniques.

Practice

The adrenaline reduces each time you present, get more comfortable, get more comfortable with it, and feel more in control. You will also get a feel for what works well and what does not regarding the spoken word.

Positive adrenalin

The trick with presentations is to turn negative adrenaline into positive adrenaline. A little adrenaline is good and will allow you to deliver a better presentation.

An Effective Presentation

Some years ago I was managing a shutdown of a North Sea platform for which we manned up to over four hundred personnel using a Flotel. As shutdown coordinator, I was required to brief the entire four hundred plus construction team offshore in the Flotels large cinema.

The presentation was aimed at the construction team and what the scope would mean to them. I covered the reasons for the shutdown and technical detail at a fairly high level before spending more time on the detailed construction worksites and activities. I relayed it as a story and used a lot of visually descriptive phrases. At one point, whilst explaining a complex crane lift, I looked up, as if to see the crane and lift. As one, the entire audience also looked up as if to see the crane and lift, thus, the audience was engaged!

Presentation Skills Summary

Depending on the current Oil price, it can be a "tight" competitive marketplace for work for Project Engineers. Imagine two candidates for a job with equal capability in experience and qualifications, but one is significantly better at presentations. Who would you hire? A small minority are natural presenters, but for the rest of us it is a skill that can be practised and developed to appear natural.

For those who wish to really develop and polish communication skills, there is a worldwide organisation called Toastmasters International[Ref6] that is dedicated to public speaking and communication, allowing for learning and practicing of public speaking and communication skills at your own pace. Meetings are held in many cities and major towns throughout the world. Try it. You will be pleasantly surprised by your improvement.

"There are only two types of speakers in the world, the nervous and the liar"
– Mark Twain

CHAPTER 8

Project Psychology

Why on earth do we have a chapter on Project Psychology? Are we not designing, procuring and constructing equipment and process plants? Yes, we are, but we are also managing that most fickle of resources: People.

Psychology is defined as the scientific study of the human mind and its functions, especially those affecting behaviour within a given context. This chapter covers some of the psychology basics and hard-won insights that can be applied to projects.

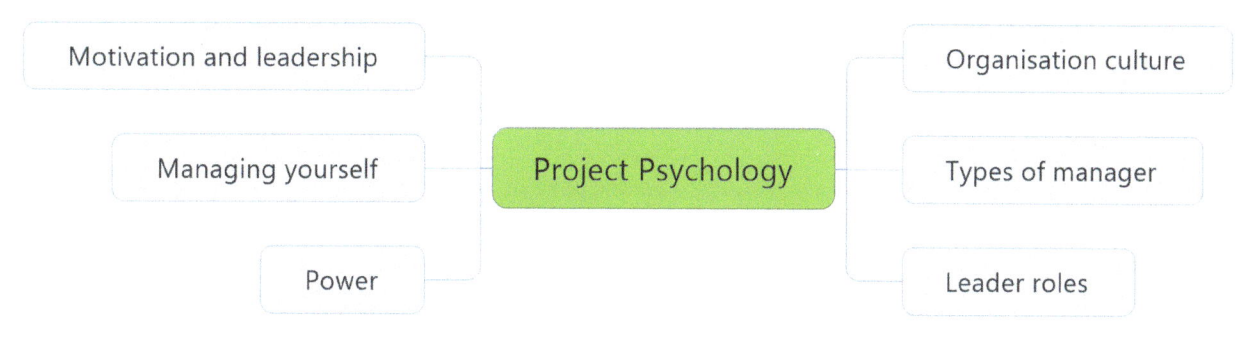

Figure 8.1 Project psychology Mind Map.

Organisation Culture

Cultures can vary significantly between companies and even within different projects of the same organisation, so there are acceptable and unacceptable behaviours within each project. For example, some projects adhere to strict meeting etiquette regarding on-time attendance and refraining from non-meeting work on phones or laptops, while others permit part-time attendees and completing non-meeting work during the meeting.

Written and unwritten rules

There will be both written and unwritten rules on what is acceptable. Learning to quickly and seamlessly fit into a project culture is an important skill for a Project Engineer. The hardest thing to change within an organisation is the culture, which takes planning and prolonged effort at all levels to achieve.

The psychological contract

In addition to the formal agreed contract of employment, there will be an unwritten psychological contract between yourself and the various team members on expected behaviours. For example, your manager might expect you to bring issues to his/her attention only after you worked on the problem and have a recommended solution for review. In return, you might expect your management to support you in front of clients and support your decisions. There will then be numerous different psychological contacts between you and your team members.

Types of Manager

Unfortunately, there is no one style that fits all in Projects and you will have to develop the ability to move between various styles depending on:

- Project culture
- Phase of the project
- Team experience
- Your own personality

At times you will need to be an information disseminator, a facilitator, a decision maker, a spokesperson, a negotiator, an organiser, etc. There are many management theories, beyond the scope of this book, however, if you are interested in them you may wish to explore the theories of Fayol[Ref 8] and Mintzberg[Ref 9], summarised below.

Fayol

In 1916, Fayol defined five functions of management that are still relevant to organisations today. These five functions focus on the relationship between personnel and management, providing points of reference so that problems can be solved in a creative manner.

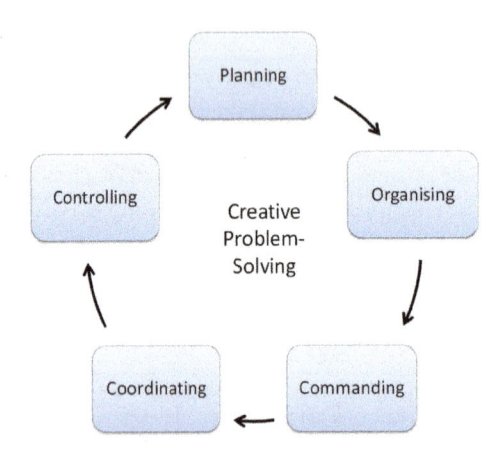

Figure 8.2 Fayol five functions of management.

Minzberg

In his 1990 book, Henry Mintzberg describes the operational work of managers in terms of ten managerial roles. Individual and situational factors influence how the Mintzberg Managerial Roles are carried out.

- Directing subordinates.
- Attending meetings as a liaison.
- Representing the organisation.
- Transmitting information.
- Analysing information.
- Allocating of resources.
- Negotiating resources.
- Problem-solving.
- Developing new ideas.
- Promoting the interests of the organisation.

My own personal style is to help the team be more effective. Don't be that demanding "table thumper" that stresses the organisation needlessly. Be the "helper," ultimately gaining you much more respect and high performance from your team.

Leader Roles

As a leader it is your responsibility to carry out the following:

Identify emerging problems

Identify emerging problems and take early corrective actions.

Act as an ambassador.

Reflect the organisations expected behaviours in both internal and external interactions.

Single point contact with the client

It is common for the Project engineer to be the single point of contact with the client's representative. This is important to ensure that all instruction for scope and changes come through the Project Engineer and that you act as the conduit for all information to the wider team. You will use your judgement on what to communicate, to whom, and in what way.

Set standards

You will need to set standards on what is expected regarding work quality, response times, behaviour at meetings, respect for co-workers, and so on.

Maintaining the project vision

Take time to focus on what project success looks like, for example:

- A completed offshore accommodation module with satisfied occupants.
- The project closing-out with achieved cost and schedule targets.
- A personal letter of commendation to yourself and your team from the client's senior manager.

When under pressure in the "thick" of the project, it can be useful to review your vision of success. It is also good to work backwards from that project success vision and insert steps to help refocus the entire team.

Suggested Task: Write down what success for your project means and review it regularly.

Support the team

Provide support to the wider team in terms of advising priorities, negotiating additional resources, clarifying issues, etc.

The collective memory

A large project will have many engineers and support staff, which will likely change as the project moves through the various phases. The constant in the project will be the Project Engineer ensuring that all issues and actions are recorded and brought to the appropriate people's notice at the appropriate time.

Without this, important issues and actions will become lost or diluted. Unfortunately, not all team members may be as diligent as you and may conveniently drop things unless gently reminded via lists, spreadsheets, or databases.

Tip: Do not try to remember everything on a project. You will fry your brain! Use good systems to record all the required actions and action parties and keep your brain for what it is best at: Processing and analysing information.

Power

The following mindmap illustrates the various types of power you will see on projects and is based on the Five Forms of Power concept [Ref7] by J. French & B. Raven

Types of Power
- Coercive Power
- Reward Power
- Legitimate Power
- Referent Power
- Expert Power

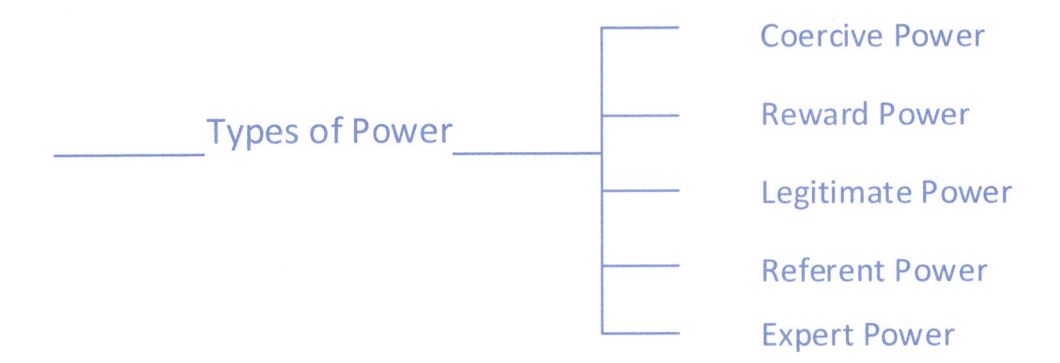

Figure 8.3 Types of power.

- Coercive: This comes from the belief that a person can punish others for noncompliance.
- Reward: This results from one person's ability to compensate another for compliance.

- Legitimate: This comes from the belief that a person has the formal right to make demands and to expect others to be compliant and obedient.
- Referent: This is the result of a person's perceived attractiveness, worthiness, and right to others' respect.
- Expert: This is based on a person's high levels of skill and knowledge.

As a Project Engineer, you will have some legitimate power, but are unlikely to have direct control over the engineering resources available. You will likely rely heavily on project management expertise (expert power) to move the team towards completion of scope.

One of the most challenging and stressful situations as a project engineer is when you are held accountable for delivery by management and clients but do not have sufficient control of the resources required to complete project deliverables. This commonly happens where engineering teams are shared across projects.

Managing Yourself

I always say that to manage your team you must first manage yourself.

Boss and worker

As Project Engineer you will likely have a large amount of autonomy about how you go about your work and schedule your weeks and days. Consider yourself as both "Boss" and "Worker". As a "Boss," plan out your week in a detached manner as you would for a subordinate, allocating actions, priorities, and dates. Then as a "Worker," simply work through the list that your "boss" has given you without challenge, argument, or jumping to the easier tasks.

Eating the elephant

As the saying goes, "How do you eat an elephant? One bite at a time."
Apply this to a project or part of a larger project. It is common, at times, to feel overwhelmed by what is in front of you, so get into "Boss" mode again to objectively cut the tasks into bite-sized pieces and delegate what you can. The remainder that you will do yourself will now become more manageable. Make a list of all these activities and highlight each item when completed.

Emotional intelligence

Emotional intelligence is the ability to identify and manage your own emotions and the emotions of others. It generally includes three skills:

- The ability to harness emotions and apply them to tasks like thinking and problem-solving.
- The ability to manage emotions, including regulating your own.
- Cheering up or calming down other people.

Many people in management positions do not have a high emotional intelligence. The good news is that it is a skill that can be learned and developed. A book I recommend is *Emotional Intelligence* by Daniel Goleman.

Example of a lack of Emotional Intelligence

A computer systems analyst I know, Fay H., is diligent, hard-working, productive, and known to deliver. Her boss asked her to complete a project that Fay estimated based on her experiences to take 5 weeks. Her boss then bluntly and quite aggressively told Fay, that it must be completed in 2 weeks. The ever-obliging Fay, however, could not meet the imposed deadline under such extreme pressure and the result was stress-induced work and absence of many weeks. Fay stated that she did not think her boss did it on purpose. Fay's boss had low emotional intelligence and would benefit from coaching and training in emotional intelligence.

Work- life balance

Managing projects can be described as similar to running endless "end-to- end" marathons with numerous flat-out sprints thrown in just for fun. To keep up the pace, you need to be physically and mentally "project fit", which means that you need to have the training, tools, and experience for the level in which you are working. You also need to have balance in your personal life and spend time on relationships, family, relaxation, and hobbies.

Tip: The most powerful tool in achieving work-life balance is a red pen. When overwhelmed with work, take the red pen and delete or cancel the non-essential activities from your diary, allowing you to "bubble up" to the surface again and regain strength.

Distractions form work

All work and no play... Find something you enjoy as a distraction from work and work colleagues. It doesn't matter what it is: singing in a choir, woodworking, cycling, etc. It will give your brain a chance to rest and will help you from stagnating.

Motivation and Leadership

Positivity engenders positivity

As a Project Engineer, endeavour to be as positive as you can. Praise, support, and exude a "can do" attitude. This positivity helps to explain why companies like Google, Yahoo!, and Virgin cultivate work environments that help their employees experience positive emotions on a regular basis. As Richard Branson said, "More than any other element, fun is the secret of Virgin's success." This isn't because fun is fun per se. It is because fun also leads to bottom-line results.

The benefits of a positive attitude can be measured and include:

- Decrease in stress and stronger immune systems.
- Increase in team performance.
- Increase in customer satisfaction.
- Improvement in cognitive abilities such as creativity and flexibility.

You deliver

The ultimate accolade for a Project Engineer is "You deliver!" This is also a tool you can use for your team members. When someone consistently delivers tell them within earshot of others, "Thank you, you delivered!"

Work smarter not harder

My company motto is "Helping you work smarter not harder." As Project Engineers, we are all normally working at capacity all the time. Trying to work harder for an extended period will result in de-motivation, burnout, and potentially illness. To be more effective, we need to be smart. The tools and techniques in this book are a baseline for you to develop and add new tools as the need arises.

Achievement moral boosters

I still do this at the start of a new project or when there are seemingly overwhelming challenges.

Write a list of all the personal achievements you have in your life such as:

- University degree
- Sports champion
- Still married after 10 years
- Project X delivered on schedule
- Article published in the company newsletter

The purpose of this is to remind yourself that no matter how bad or challenging the current situation feels, you are not a failure but an achiever. It helps you feel good, which is the objective.

Application of sports psychology

Sports psychology has applications in projects. Growing up, I had no interest in soccer, but throughout my Project Management career I have become very interested in the management viewpoint of football, the psychology that football managers use to deliver a winning team.

There are 5 common characteristics between sport and projects

- Goals: The Team/Project need clear identifiable goals to aim for.
- Feedback: Detailed constructive feedback after events is important as it is by feedback that we improve.
- Rules: We need to know the rules of the game (project procedures). We all need to know what is permitted and what is not.
- Support: We need to support all members of the team both in workload and emotionally.
- Roles and Responsibilities: Everyone needs to know their R&Rs so that we do not miss of duplicate activities.

The following table illustrates the typical differences between sport and a project.

Sport and projects comparisons

Characteristic	Sport	Project
Feedback	Immediate	Delayed
Rules	Well-defined and understood	Undefined, changing, and not commonly understood
Reinforcement	Immediate or very soon	Rare, non-existent, or too late
Goals	Clearly specified with easy-to-determine progress	Unclear with difficult-to-determine progress
Roles and Responsibilities	Clearly specified and well-understood	Often loosely defined and poorly understood

Table 8.4 Sports & projects comparisons.

Task: Review your project for each of the above areas and perform a gap analysis on any shortcomings. You will now have a list of actions to improve your project.

Project Psychology Summary

Smart Project Engineers know what makes them and their team tick. By understanding yourself and your team and by helping them, you will achieve the desired successes.

"The path of least resistance is the path of the loser". - H. G. Wells

What I take from this is to do the hardest things first for the most benefit.

CHAPTER 9

Stress

The objective of this chapter is to enable you to identify stress in yourselves and others, and then use the information to reduce and manage the effects of stress. But first a question, and try to answer honestly. Have you ever been affected by stress at any level? We will also ask this later on in the chapter.

Figure 9.1 Stress Mind Map.

Unchecked stress is one of the major problems of the modern world, with all the stimuli, deadlines, and demands on our time. Stress in life cannot be eliminated, but by understanding the causes and effects of stress and adopting some simple management strategies, it can be reduced to manageable levels.

What is Stress?

When you feel threatened, your nervous system responds by releasing a flood of stress hormones, including adrenaline and cortisol. Adrenaline increases your heart rate, elevates your blood pressure, and boosts energy supplies. Cortisol, the primary stress hormone, increases sugars (glucose) in the bloodstream, enhances your brain's use of glucose and increases the availability of tissue repair substances.

These hormones rouse the body for emergency action: your heart pounds faster, muscles tighten, blood pressure rises, breath quickens, and your senses become sharper. These physical changes increase your strength and stamina, increase your reaction time and enhance your focus. This is known as the "fight or flight" response and is your body's way of protecting you.

Good stress

When stress is within your comfort zone, it can help you stay focused, energetic, and alert. In emergency situations, stress can save your life, giving you extra strength to defend yourself or escape from danger. Stress can also help you rise to meet challenges. Good stress is what keeps you on your toes during a presentation at work. But once it goes beyond your comfort zone, stress stops being helpful and can instead cause major damage to your mind and body.

Bad stress

Repeatedly experiencing the "fight or flight" stress response in your daily life can lead to serious health problems. Chronic stress disrupts nearly every system in your body, impacting your immune system, upsetting your digestive system, raising your blood pressure, increasing the risk of heart attack and stroke, and affecting many other physical and mental factors. This is bad stress.

Stressors

The situations and pressures that cause stress are termed stressors. We usually think of stressors as being negative, such as an exhausting work schedule or a rocky relationship. However, anything that puts high demands on you can be stressful, including positive events such as getting married, buying a house, going to college, or receiving a promotion.

Events and situations can be different for different people. What is negatively stressful for one person may be a good challenge for someone else.

Consider the following examples;

- Karen is terrified of getting up in front of people to perform or speak, while her best friend, Nina, lives for the spotlight.

- Phil thrives under pressure and performs best when he has a tight deadline, while his co-worker, Matt, shuts down when work demands escalate.

Exercise: Write down your major sources of stress. These will be different for everyone. When you recognize them, you will be able to manage them.

Three Types of Stress

Stress can be divided into three generally-accepted categories:

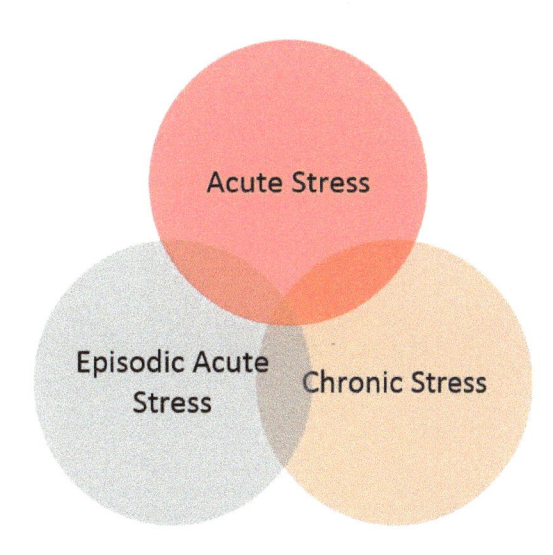

Figure 9.2 Types of stress.

Acute stress

Acute stress is the most common type of stress and is your body's immediate reaction to a new challenge, event, or demand, triggering your "fight or flight" response. Examples include a near-miss motorcar accident, an argument with a family member, or a costly mistake at work. In response, your body turns on this biological response. This type of stress starts and ends quickly.

Episodic acute stress

Episodic acute stress is acute stress that happens frequently. "Worriers" or pessimistic people who tend to see the negative side of everything tend to have episodic acute stress. These people always seem to be having a crisis, and are often short-tempered, irritable, and anxious. Episodic stress is not like chronic stress, though, because this type of stress ceases from time to time.

Chronic stress

If acute stress isn't resolved and begins to increase or lasts for long periods of time, it becomes chronic stress. This type of stress is constant and doesn't go away easily. It can stem from such things as:

- Financial worries.
- Family issues.
- An unhappy marriage.
- A bad job.
- Just being too busy.

Chronic stress can be detrimental to your health as it can contribute to several serious diseases and health risks covering both physical and mental health.

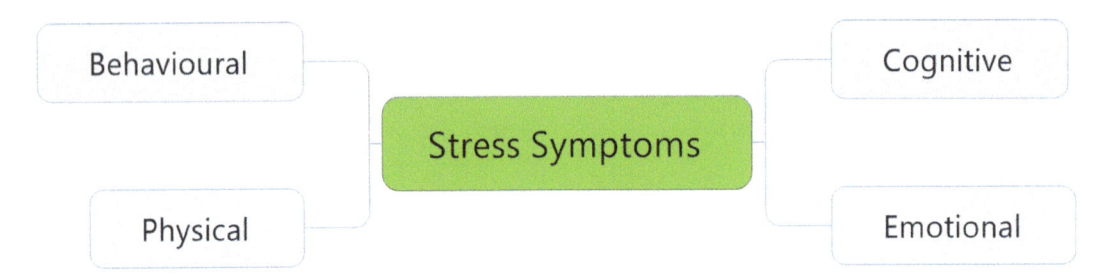

Figure 9.3 Stress symptoms.

While researching this chapter I was surprised at the number and breadth of symptoms listed below that can be attributed to stress.

Cognitive symptoms

- Memory problems.
- Inability to concentrate.
- Poor judgment.
- Seeing only the negative.
- Anxious or racing thoughts.
- Constant worrying.

Emotional symptoms

- Depression or general unhappiness.
- Anxiety and agitation.
- Moodiness, irritability, or anger.
- Feeling overwhelmed.
- Loneliness and isolation.
- Other mental or emotional health problems.

Physical symptoms

- Aches and pains.
- Diarrhoea or constipation.
- Nausea, dizziness.
- Chest pain, rapid heartbeat.
- Loss of sex drive.
- Frequent colds or flu.

Behavioural symptoms

- Eating more or less.
- Sleeping too much or too little.
- Withdrawing from others.
- Procrastinating or neglecting responsibilities.
- Using alcohol, cigarettes, or drugs to relax.
- Nervous habits (e.g. nail biting, pacing).

If you answered "no" to the question at the beginning of this chapter, please review the above list of stress symptoms and re-consider your response.

Over the years with the highs and lows of many Oil & Gas projects, I have experienced quite a few of the aforementioned symptoms, but with some research into stress reduction and trial-and-error, I am now rarely stressed out to the point of negatively impacting my work or personal lives.

Project symptoms

The main symptoms of a stressed workforce will show itself as decreased morale and productivity, hence it should be addressed promptly. A stressed project team will not perform as well as possible and many will leave as soon as they have an opportunity.

Managing Stress

Stress affects each person differently. Some people may get headaches or stomachaches, while others may lose sleep or get depressed or angry. People under constant stress tend to get sick a lot. Managing stress is important for staying healthy.

It is impossible to completely get rid of stress, so the goal of stress management is to identify your stressors and apply techniques and mitigations to reduce the negative stress that is induced.

Simple tips

- Take care of yourself, by eating healthy, exercising, and getting plenty of sleep.
- Find support by talking to other people to express yourself.
- Connect socially, as it is easy to isolate yourself after a stressful event.
- Take a break from the cause of stress.
- Avoid drugs and alcohol, which may seem to help with stress in the short term, but can actually cause more problems in the long term.

Work-Related Stress

Work-related stress is prevalent in many organisations and industries for many reasons. It is underreported and under-discussed. Here are some practical tips:

- Take a walk, swim, run, do, yoga, meditate, etc… at lunchtime

- List all the tasks you have been asked or need to do and estimate the percentage of workload. If it's more than 100%, delegate the tasks downwards, across, or upwards.
- If there is too much work on your plate, let your boss know since it is also his/her problem.
- Reduced work hours/days in the week.
- Seek an internal transfer if stress is still an issue.
- If stress remains an issue, visit your doctor. If unresolved, leave your work environment! Your health is more important!

Stress management is all about looking after yourself, not running yourself too hard but instead putting into place some simple measures to reduce and manage said stress.

Stress in Colleagues

It has been estimated that stress in the workplace is an issue for almost 40% of office workers. As a Project Engineer, it is in your project's interest to minimise stress in your team. Also, as a human being, it is not exactly nice to see team members stressed out and unhappy. It is usually fairly easy to see when someone is stressed out so as a Project Leader you should make it part of your responsibilities to assist in reducing stress whenever possible in the workplace. The following methods can be used to assist your team members:

Listen to the stressed team member

Meet with the team member informally and inform them that you have noticed they are stressed. Create an environment for them to express themselves, and offer to help if possible. Often, just listening to your team members lets them know that you can appreciate their situations and are willing to help.

Identify the source of the problem

There are three main stressors in a working environment:

- High workload.
- Uncertainty on how to proceed.
- Interpersonal conflicts.

High workload

Assist the team member in establishing his/her workload and reducing it to a manageable level. This may require meetings and agreements with other managers. For very stressed individuals, you may have to reduce workload well below their normal capacity for a short time to allow them to recuperate.

Uncertainty on how to proceed

Not knowing the next steps can lead to stress, so it is important to use your delegation skills, provide clear guidance, and support the team.

Inter-personal conflict

People can be a source of stress, as discussed in the conflict management chapter. The first step is to see if the parties can sort out the conflict amongst themselves. If not, then provide support to the team member/s whilst implementing additional conflict management techniques.

Stressing the Business

Running Projects can be stressful in and of itself, but unfortunately there are people at all levels who can artificially "stress the business." These people shout, bully, intimidate and make unreasonable demands. In all cases, you should be able to apply conflict management techniques, remain polite and professional, and remember the following advice I was given many years ago: *"Peter, it is only work."*

Stress Summary

Half the battle of dealing with stress is awareness that there is stress in yourself and others. Not all stress is negative, and in fact can be turned into a positive stressor, which is the biggest win, such as with public speaking. Stress affects all of us on varying levels, whether we admit it or not. There are many books and techniques available on managing stress. If you are affected, try out different techniques and find out what works for you.

"One thing is certain: our families are important. Don't get so stressed out and so pre-occupied that we neglect one of the greatest things that Life has given us, and that's our families."- Victoria Osteen.

CHAPTER 10

The Position Statement

Many projects can become confused and muddled, especially those with urgent scopes, rapidly evolving definitions and little to no control. This adds easily avoidable costs, delays, and stress to the project itself and the project team. The following Position Statement methodology is a proven, structured method of controlling the workscope and coordinating the activities of both the internal team and wider stakeholders.

Figure 10.1 Position Statement mindmap.

Objective of the Position Statement

In my experiences, many urgent and fast-track scopes are typified in long e-mail chains that are distributed to a wide audience in which key points and actions are unclear and buried in the text, and meetings are often too free-flowing with inadequate minutes taken.

The objective of the Position Statement is to provide clear project status that allows efficient and effective coordination of all stakeholders.

Initial Generation of the Position Statement

As the Project Engineer, it is incumbent on you to clarify and maintain the scope definition. Start by collating all the available information relating to the scope, such as e-mails, reports, meeting minutes, discussions with stakeholders, and so on. Review the input information and generate a draft document of your understanding of the

- Problem definition.
- Scope requirements.
- Schedule requirements.
- Internal and external stakeholders
- Next steps and actions.

Armed with this first draft, you will then review it with your internal project team. Figure 10.1 shows the iterative process of generation and update.

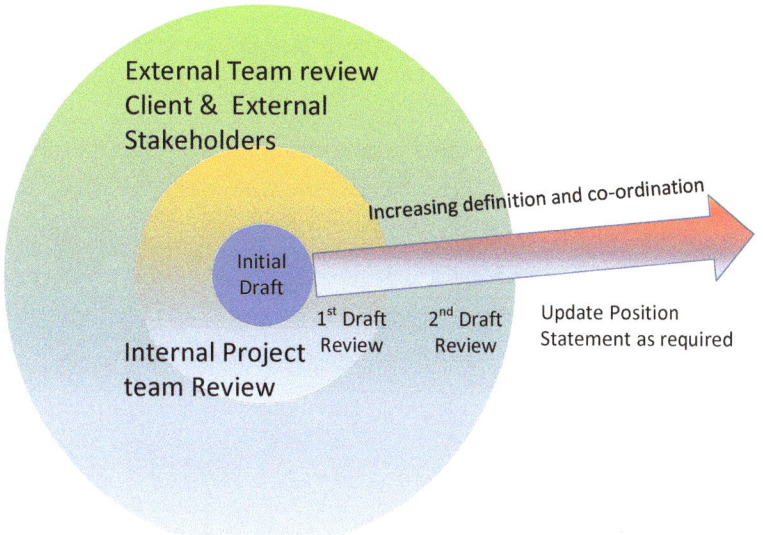

Figure 10.2 Position statement process.

1st Draft Review

You now have an internal baseline document that you can review with the entire internal team, which will provide additional information and clarifications. At the end of this review, your internal team should have achieved consensus on the issues, scope definition, and next actions. The next step is to widen out the consensus of the project team to the external stakeholders.

2nd Draft Review

This step involves reviewing the position statement and scope with external stakeholders, who might include client, subcontractors, and certifying authorities. At the end of this review, you should have gained accurate information on the issue and the proposed scope in a concise format. During face-to-face meetings or conference calls, walk the external stakeholders through the position statement to obtain their input and clarification. At the end of this stage, all the external and internal stakeholders should be on the same page regarding project scope and next actions.

Tip: Remember the team and stakeholders are busy people who do not have time to wade through long e-mail chains and potentially misinterpreted requirements.

Final Document

Further iteration

The position statement should have a date and revision at each issue. For ease and speed of assimilation, I suggest that changes from the previous versions are highlighted in a different colour.

On very urgent and fast-moving scopes, it would not be unreasonable to update the position statement daily, reducing frequency as the scope definition settles down.

Benefits of a position statement

I have been using the position statement methodology for years and have coached many Project Engineers on its uses as an effective and time-saving tool. The position statement:

- Provides a clear unambiguous scope baseline at a point in time.
- Can be used as the basis of a more detailed scope document.
- Can be used as the basis for formal change control.
- Can be revised as scope changes.
- Assists in managing challenging clients.

Duration to complete position statement

The preparation of a position statement is an investment that will save time and reduce potential confusion, especially from those further from the centre. Typically, once input information has been reviewed, the position statement can be completed within an hour. As Project Engineer, you must perform the information review legwork for the entire team.

Typical sections

Typical sections of a position statement are shown below, although I recommend you modify them to suit your particular scope:

- Background on the issue.
- Work completed to-date.
- Identified solutions.
- Issues requiring further discussion.
- Single point contacts.
- Next steps.

Position Statement Summary

Using a position statement brings all the stakeholders to the same place. It is relatively quick and easy to do, and it reduces stress on the organisation as a whole. Try it with your next scope!

"Sometimes when co-ordinating stakeholders, it feels like you are trying to herd cats."

CHAPTER 11

Planning

Successful planning of a project involves a range of techniques to allow you to build a plan, gain valuable insights from that plan, and take corrective actions to keep your projects on schedule. The following techniques will allow you to both carry out strategic top-down planning for significant projects, short-term detailed planning for short duration key activities, and to effectively use your organisation's planning and scheduling documentation.

Figure 11.1 Planning Mind Map.

Where to start?

Although you are likely to have a planning department to assist in developing and maintaining a plan, as the responsible Project Engineer, you own the plan and you will need to provide the initial input to the planner. This is normally formally communicated at the project kick-off meeting. Once you have prepared the "position statement" and with the help of the entire team, you should be able to answer the following questions:

- What is the client's construction completion requirement date?
- How long will construction take?
- How long will it take to procure the long lead items?
- How many construction workpacks will be required?
- How long will engineering take?
- When do you anticipate approval to formally kick-off and assign resources?
- Are there any key milestones, such as facility shutdowns, to take into account?

CHAPTER 11 PLANNING

If your organisation has a formal kick-off procedure, you will likely have a kick-off planning checklist that will cover the key planning input questions. Do not worry about being too accurate at this stage as planning is iterative and both accuracy and detail will improve with each review.

Tip: Start at the end with the required completion date and work backwards. This is known as right-to-left planning.

Level 1 Logic and Schedule

After the kick-off, you will now have enough input information to start building an initial top-down plan/logic.

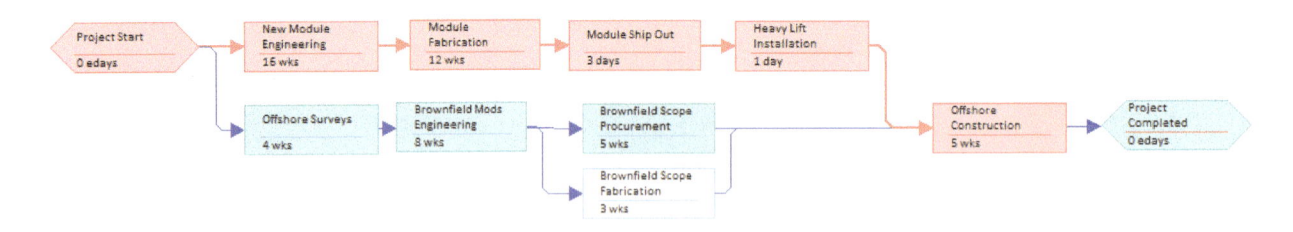

Figure 11.2 Simple level 1 logic.

I find building these Level 1 (L1) logic diagrams are easiest by printing out all the activities and using traditional paper-based cutting and pasting to properly sequence. This will usually go through a number of iterations and can be used to explain your understanding of the planning sequence to the wider team and gather their input. The exercise of working this logic through with the team is probably as valuable as the final plan as it coordinates the team's input.

After developing the Level 1 logic you can produce a draft schedule based on the logic (see Figure 11.3).

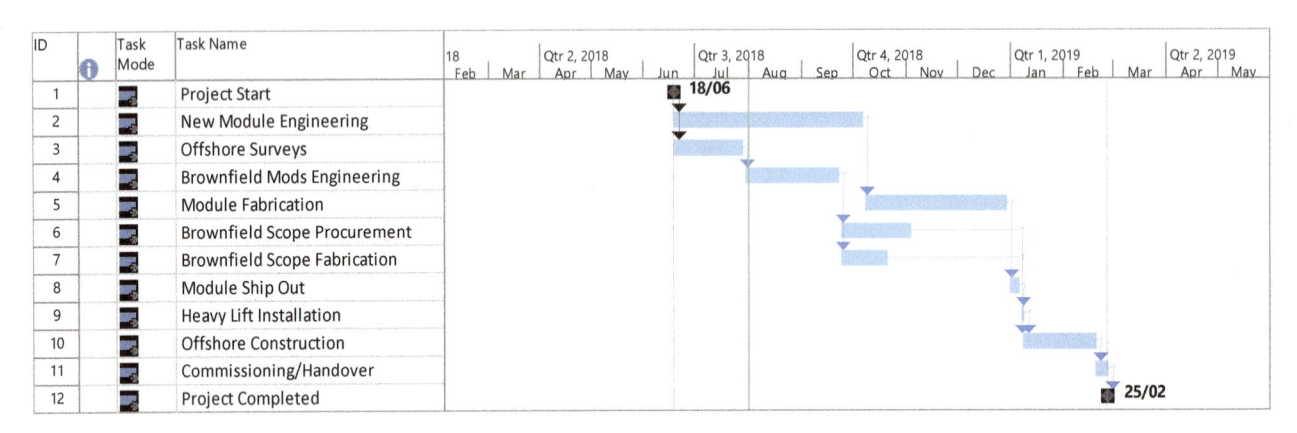

Figure 11.3 Simple level 1 schedule.

The top down strategic L1 logic and schedule can be used as a communication tool to the wider project team and as input for the project planner to create the detailed plan. Level 1 logic is also useful during the construction and commission phase to show relative completion of the project.

I have prepared wall posters on numerous projects showing the construction and commissioning sequence of all the main activities. This wallchart is then marked up regularly with a single slash indicating activity started, and a cross indicating that the activity is complete. They are great for briefing management and people who are new to the project, such as specialist subcontractors.

Tip: On multi-organisation/company projects, add key contractor logos to the graphics to help them feel like part of the wider project team.

The Project Critical Path

Many people incorrectly use the term "critical path," assuming it is their interpretation or opinion of the most important or urgent activities of a project. Simply put the critical path is the longest duration path of activities through a network of activities, and should be the shortest possible time that a project can be completed in. Please see Figure 11.2 above for the critical path items marked in red

The critical path method (CPM), is used widely today in planning and is based on a project modelling technique developed in the late 1950s by R. Walker of DuPont and J. Kelley Jr. of Remington Rand[Ref8].

The key features of the CPM are as follows:

- A list of all activities required to complete the project (typically categorised within a work breakdown structure).
- The time (duration) that each activity requires to complete.
- The logical dependencies between the activities.
- Logical end points, such as milestones or deliverable items.

Using these values, CPM calculates the longest path of planned activities to logical end points, or to the end of the project, as well as the earliest and latest that each activity can start and finish without increasing the overall project timeline. This process determines the longest path through the plan and which activities are "critical".

Precautions when using critical path method

There may be multiple "near critical paths" that can become critical if durations alter even slightly. Near critical paths indicate risks that must be managed to ensure they do not become critical.

Project Engineers "9 Bar Blues"

The name of this technique has a loose, visual link to a short music score of a few lines.

"9 Bar Blues"
Passive heave Compensation Offshore Implementation

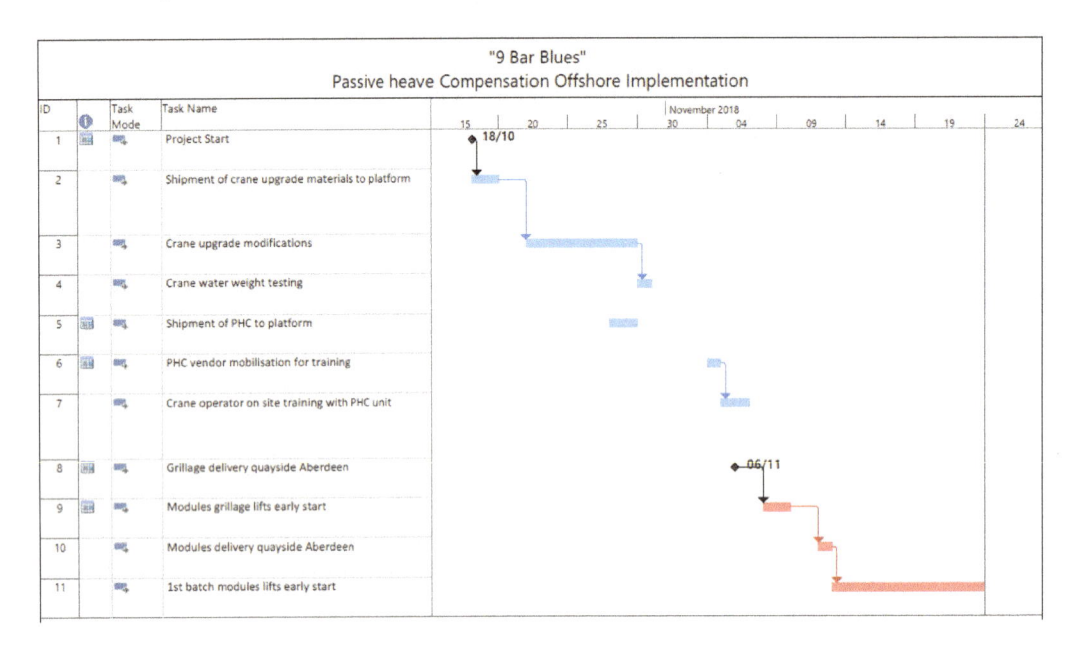

ID	Task Mode	Task Name
1		Project Start
2		Shipment of crane upgrade materials to platform
3		Crane upgrade modifications
4		Crane water weight testing
5		Shipment of PHC to platform
6		PHC vendor mobilisation for training
7		Crane operator on site training with PHC unit
8		Grillage delivery quayside Aberdeen
9		Modules grillage lifts early start
10		Modules delivery quayside Aberdeen
11		1st batch modules lifts early start

Figure 11.4 Project Engineer "9 Bar Blues".

This is what I call the "thinking paper" where the Project Engineer roughs out a draft plan to allow him/her to visualise the activities, their individual durations, and the sequence of implementation. It is normally used prior to having the planning department produce a detailed schedule or when short-term scope management is required.

The 9 bar blues can be produced in minutes, drawn freehand, and are great communication tools. I often use 9 bar blues schedules as agenda items for meetings, discussing each bar in turn, generating actions, and adjusting dates and progress. This is particularly effective for coordinating multiple subcontractors and stakeholders via conference call meetings.

Short-term schedules

Short-term schedules are often required where there are daily or even hourly updates. Planning departments normally cannot respond to fast turnaround times and the PE must prepare and update short-term schedules. On a significant platform and subsea field shutdown I managed onsite some years ago, I provided a detailed hourly schedule update every 12 hours that was then e-mailed to the drilling rig and diving support vessel supporting the shutdown operations.

Software for planning

Large organisations typically use a complex interconnected planning software system that link to data streams from various parts of the organisation.

In my experience the planning software that is essential for scheduling, reporting progress and understanding resource requirements is not nimble enough for short-term planning. For project engineers, planning "thinking paper", I find Microsoft Project or similar to be useful. It is relatively easy to learn how to use and output clear and simple schedules for communication purposes. Microsoft Project can also produce logic networks also known by the acronym PERT that show the logical relationship between activities.

Project Work Breakdown Structure

The project work breakdown structure (WBS) is a hierarchical structure that breaks projects down into smaller activities to authorise, control, and report on in manageable pieces. Your organisation may have procedures defining what each level consists of, or you might define it specifically for your particular project.

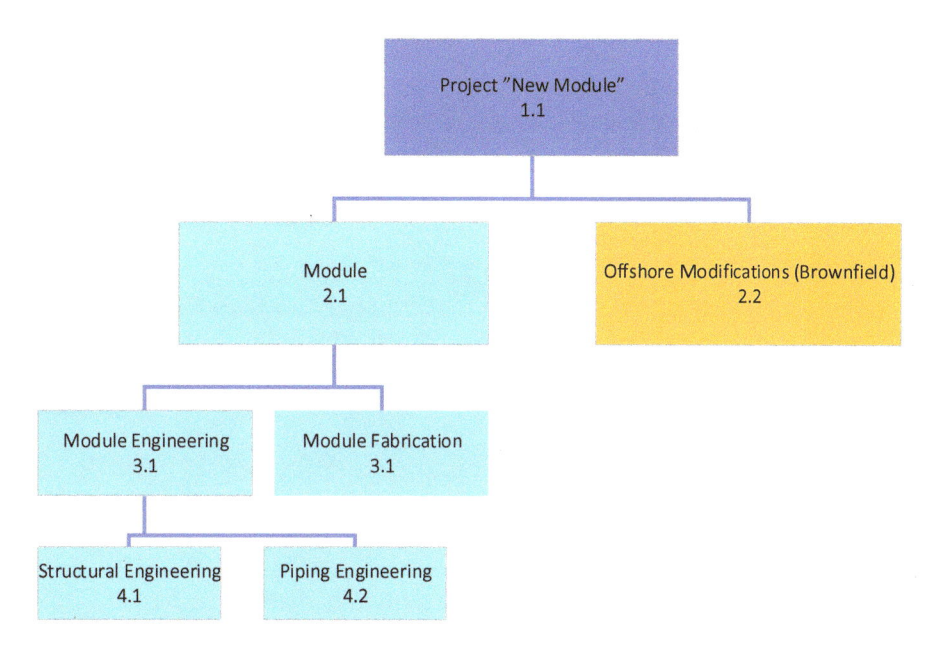

Figure 11.5 Work breakdown structure example.

The beauty of a WBS is that different work packages can be assigned to different people or subcontractors, ultimately being sequenced and combined into the project plan.

Best practices for WBS

The WBS is one of the main foundations of a project hence is vital that time is spent developing and communicating it. Planning, cost management, and reporting will depend on it.

WBS dictionary

Each WBS element should be concisely and fully defined to detail its content rather than to describe in a few words. You can avoid scope creep by defining what will be done and what will be excluded for each element.

Common planning system

One of the best practices is when the client or main contractor generates and maintains the upper levels of the WBS with each of the subcontractors responsible for the lower levels using a subset of the higher level plan and the same software system and databases.

This must be set up early in the project, and when achieved, it helps to avoid:

- Multiple different plans.
- Manual data transfer errors.
- Timing errors between plans.

Bottom-Up Planning

We have explained top-down and strategic planning and how they work. The other common process is "Bottom Up" planning which is typically carried out at WBS levels 3 or 4, i.e. engineering Cost Time Resource (CTR) deliverable and workpack or jobcard Bottom-Up planning is split into the engineering and construction phases.

Engineering cost time resource sheets

When the estimate is being compiled for a project, the engineering team submits CTRs that specify the engineering deliverables they will complete for a particular scope, such as:

- Offshore surveys.
- Reports.
- Calculations drawings.
- Requisitions.

There are normed (see Chapter 12) hours associated with each of these activities that will be part of the agreed contract. The Project Engineer will collate and review all the CTRs. This review will include challenging what the disciplines have estimated and reducing or increasing, as required, to make the estimate as accurate as possible.

Tip: Some engineers will put in overly high estimates. Get to know who they are, why they put high estimates, and challenge accordingly.

Planning will sequence the engineering activities, produce a draft schedule for review and comment, and a draft manning histogram showing the numbers of engineers and designers and when they are required. There will normally be peaks for periods where high levels of a specific discipline of engineering or resource is required. The planner will then smooth the resource peaks to a manageable profile as resources are not infinite and immediately available.

A workable plan will be baselined when it has been generated and accepted by the discipline leads and project management teams. Baselining freezes what is planned and allows for progress of each planned activity to be measured and monitored, thus allowing for corrective action to be taken when an activity starts to fall behind schedule.

The level 4 plan is a detailed plan which can be printed out in many formats and filters, such as:

- By milestone.
- By discipline.
- By late activity.
- By critical activities.

Key performance indicators

Project performance payments are often linked to schedule targets called key performance indicators (KPIs). These KPIs are target dates that can trigger pre-agreed enhanced payments when achieved. KPIs often fall on a sliding scale, with maximum profit achieved when the earliest deadlines are met, and minimum or no profit when the latest permissible dates missed. When carefully selected, KPIs should align both the client and the contactor to common objectives.

How to Use the Planning Information

Project engineers can be swamped with schedules in terms of both variety and quantity. As previously mentioned, there are many filters on how a plan can be produced for different end users. For a Project Engineer working on a particular project or part of a project in the engineering and workpack phase, a filter by project and discipline is probably the most useful. This filter can be used to do a number of things:

Front lining plan

Front lining the project plan allows you to determine which activities are falling behind schedule. This is sometimes done as part of planning preparation or by drawing a red line down the "Time now" date. Anything behind the planned progress should be highlighted for discussion with the appropriate disciplines.

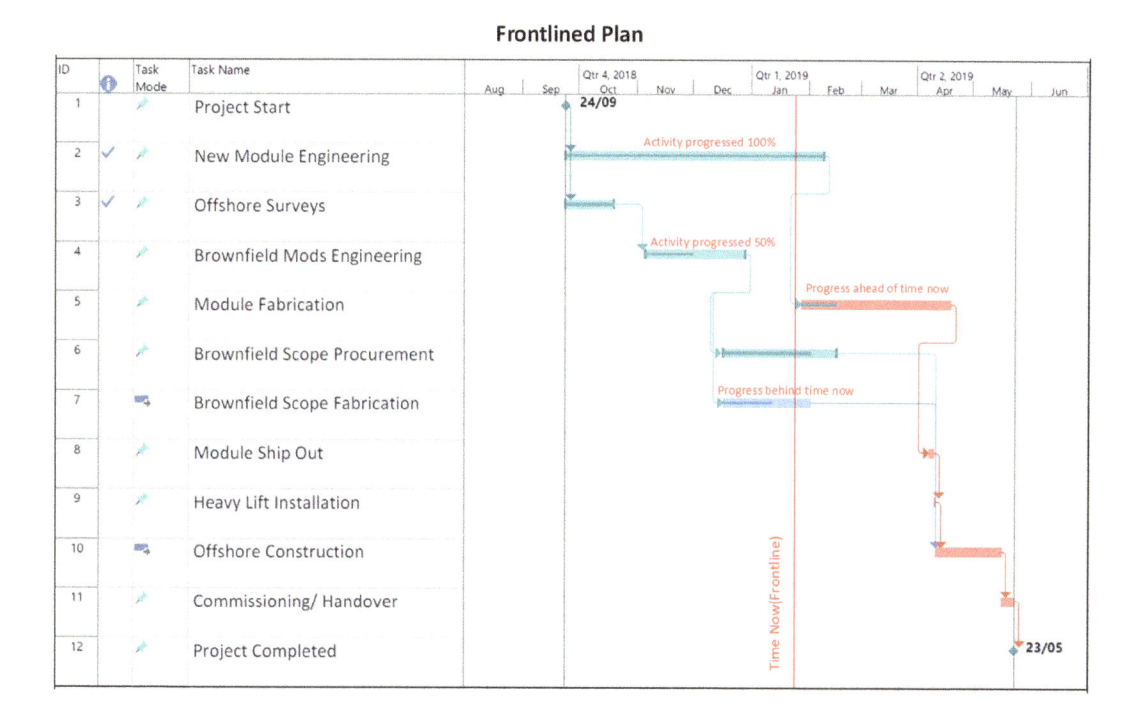

Figure 11.6 Front-lined plan.

Plan as a progress meeting agenda

Front lined plans are often used as agendas for engineering discipline project planning meetings. With all the relevant disciplines represented, you and your planner will typically run through each line of the plan, asking for reasons an activity might be behind schedule and then agreeing on the required corrective action to recover schedule or minimise slippage. If completed on a weekly basis, you will become intimately familiar with the plan as well as a solid understanding of the problem areas, people and issues.

Tip: Many engineering disciplines dislike these meetings because they are being held accountable. Assist in removing blockers and bottlenecks where possible, but hold the team member accountable as you are also accountable to your Project Manager or client should there be any slippage.

Plan types and ownership of plans

In client operations and projects there are three main categories and timescales for planning:

Long-Term Plans: Up to 5 years in the future sets long-term targets for the asset and are typically maintained by a client planner at a fairly high level. These targets are used to communicate and report against strategic objectives.

Medium Term plans: Duration of circa 4 weeks up to about a year for the duration of a project. These will be jointly maintained by the project and client planners. Medium-term plans will be more detailed with specific scope activities.

Short-Term Plans: Short term plans can be 2 weeks or less and are typically owned by offshore operations. They will be reviewed on a weekly basis between offshore and onshore management teams wherein material, manpower availability, and priorities are decided by offshore operations.

Plan entry criteria

Due to the complex logistics associated with mobilising men and equipment for offshore construction activities, as well as the rigorous permit requirements to carry out any offshore work, most operators have developed plan entry criteria, which is a series of "gates" wherein activities cannot be entered into the plan without complying at typically 90-days, 28-days and 7-days. Only in exceptional situations and with client management agreement are scopes allowed to proceed forward without fully meeting the gate criteria.

Planned Activity Criteria	Responsible	90 Day (Gate Keeper:Operations Manager)	28 Day (Gate Keeper: OIM)	7 Day (Gate Keeper: OIM)
Financial approval	Operations Manager	Circulated for review	Approved	Approved
Workpack	Sponsor	Defined & on schedule	Distributed offshore	Available onsite
POB: Personnel on Board/ Vendor/resources	Sponsor	Contacted and agreed	Flights/beds confirmed	Mobilisation dates confirmed
Platform resources	Sponsor	Identified	Platform work order approved	Platform work order scheduled
Materials/Equipment	Sponsor	Identified & on schedule	Materials available onshore	Materials available offshore
SIMOPS Simultaneous Operations	Sponsor	Identified	SIMOPS agreed	SIMOPS validated
Over-side work	Sponsor	Identified	Vessel plan in place	Plan verified
Access scaffold	Sponsor	Identified	Scaffold work order approved	Scaffold work order issued onsite
Lifting Plan	Operations Manager	Defined & on schedule	Finalised	Approved and available onsite

Table 11.7 Simplified plan entry criteria.

Planning Progress Reports

Planning progress reports are a valuable tool for establishing progress and providing early warnings on issues that may require further investigation.

Please reference the following engineering progress report from a project of some 11,000 manhours (circa 6 man-years).

It is recommended that you review this report prior to the disciplines progress meeting and prepare the following:

- Check the project's current budget (in manhours) versus forecast at completion. If the forecast is above budget, you will need to find out which disciplines are over and why.
- Look down the disciplines in the variance column to determine which disciplines are forecasting to overspend. In discussion with the disciplines, determine why they are overspending, determine if the additional work is approved,

CHAPTER 11 PLANNING

is there preferential engineering, or is the workload underestimated etc. You, as PE, will be required to explain any requested increases to the client.

- Look at productivity-to-date ratio. Productivity is defined as earned hours divided by the actual hours. A rule of thumb is to question productivities above 1.2 or less than 0.8. A low productivity may indicate a team is not performing or scope growth/change has not been captured in a change or variation request. A high productivity may indicate an initial overestimate.

	DISCIPLINE	BASELINE BUDGET	ORIGINAL BUDGET	CURRENT BUDGET	ACTUAL % COMPLETE	ACTUALS PERIOD	ACTUALS TO-DATE	EARNED PERIOD	EARNED TO-DATE	FORECAST AT COMPLETE	FCAST REMAINING HRS WITHOUT PROD	PROD. PERIOD	PROD. TO-DATE	VARIANCE
	Example - CTR Module Engineering									Report Date:	01-Nov-2018			
	Summary Manhours - Discipline									Actual Cut-Off:	30-Oct-2018			
	Project: 17062529 - Chieftan													
122	STRUCTURAL	3,991.99	5,013.26	4,901.22	90.38	20.00	4,263.50	17.62	4,429.94	4,734.78	471.28	0.88	1.04	-166.44
128	PIPE SUPPORT	28.99	2,244.37	2,079.12	65.06	199.00	1,614.50	126.21	1,352.58	2,341.04	726.54	0.63	0.84	261.92
131	MECHANICAL	1.00	154.80	151.33	54.52	8.00	62.00	10.80	82.51	130.82	68.82	1.35	1.33	-20.51
132	PIPING	3,114.50	6,629.90	6,396.82	85.01	234.50	4,868.00	294.81	5,437.73	5,827.09	959.09	1.26	1.12	-569.73
136	METALLURGY	942.22	1,561.83	1,410.63	87.23	77.00	1,247.50	51.88	1,230.44	1,427.69	180.19	0.67	0.99	17.06
139	PIPE STRESS	24.00	301.00	308.20	55.03	47.00	172.00	19.80	169.60	310.60	138.60	0.42	0.99	2.40
140	ELECTRICAL	412.20	2,007.21	2,037.21	85.58	188.50	1,648.00	198.25	1,743.52	1,941.69	293.69	1.05	1.06	-95.52
151	INSTRUMENTS	73.73	235.29	269.95	99.26	3.00	256.50	3.00	267.95	258.50	2.00	1.00	1.04	-11.45
180	CONSTRUCTION (C3)	0.00	0.00	235.50	0.00	0.00	0.00	0.00	0.00	235.50	235.50	0.00	0.00	0.00
182	C3 - PIPING	3,984.44	4,701.84	5,684.44	98.29	43.00	5,419.00	162.16	5,587.49	5,515.95	96.95	3.77	1.03	-168.49
183	C3 - ELECTRICAL	0.00	138.40	138.40	95.95	30.00	113.00	24.80	132.80	118.60	5.80	0.83	1.18	-19.80
184	C3 - INSTRUMENTS	13,192.17	17,717.57	18,612.85	94.93	379.00	17,022.50	391.41	17,668.89	17,966.46	943.96	1.03	1.04	-646.39
185	C3 - COMPLETIONS	4,590.36	6,382.44	4,814.28	60.90	45.00	2,825.00	51.12	2,931.80	4,707.48	1,882.48	1.14	1.04	-106.80
186	C3 - PRE-COMMISSIONING / COMMISSIONING	4,370.00	4,370.00	4,370.00	35.82	50.00	1,456.00	39.50	1,565.50	4,260.50	2,804.50	0.79	1.08	-109.50
187	C3 - STRUCTURAL	21,392.94	29,432.79	29,008.92	94.88	334.00	26,672.00	234.69	27,523.46	28,157.46	1,485.46	0.70	1.03	-851.46
250	TECHNICAL SAFETY	89.70	299.86	321.86	52.12	18.50	180.50	10.00	167.74	334.62	154.12	0.54	0.93	12.76
256	STRUCTURAL ANALYSIS	301.12	341.12	341.12	99.41	0.00	323.25	0.00	339.12	325.25	2.00	0.00	1.05	-15.87
Project Total:		56,509.36	81,531.68	81,081.85	87.11	1,676.50	68,143.25	1,636.04	70,631.07	78,594.03	10,450.78	0.98	1.04	-2,487.82
Grand Total:		56,509.36	81,531.68	81,081.85	87.11	1,676.50	68,143.25	1,636.04	70,631.07	78,594.03	10,450.78	0.98	1.04	-2,487.82

Figure 11.8 Plan manhour report.

Poor period productivity can indicate missed progress reporting or a problem within a specific discipline. If you can spot these issues so will the client as PE you will be held accountable for your project's performance.

All the analysis only works if the reporting is accurate. Unfortunately, some discipline leads use engineering man-hours as money and advise their teams on which CTRs to book irrelevant of the scope. Encourage true progress reporting from disciplines whether good or bad as this will give you the most accurate picture of where the project is and any required support tor corrective actions.

S Curves/Run-Down Curves.

Wading through pages of detailed schedules, barcharts, and detailed progress reports is time-consuming, so visual planning documents are used, such as S Curves and run-down curves. These types of curves are used for presentation purposes at management and client meetings and as the Project Engineer you will be expected to explain features and trends.

S Curves

A planning S curve is a plot of planned progress against actual and forecasted progress. The name derives from the typical "S" shape of these curves.

Typical Planning S Curve

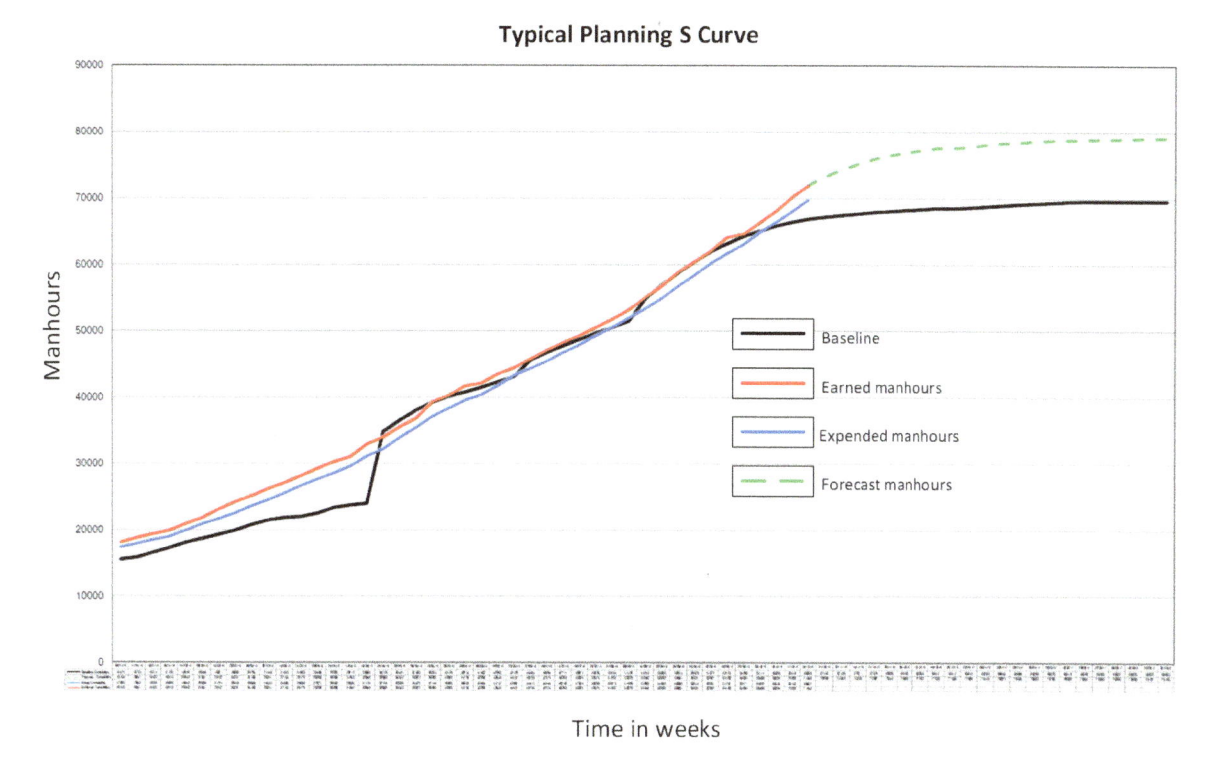

Figure 11.9 Planning S curve.

With a little practice, S curves can be quickly and easily read and the project progress and trends determined.

The curve is built from the baseline plan which shows planned progress at particular points in time. As weekly progress is reported, it is also plotted on the curve.

The curve in Figure 11.9 shows the following key features:

- The black line "step up" shows that the plan has been re-baselined due to added scope.
- Earned manhours are above expended manhours, indicating good productivity.
- The forecast above the baseline shows the increase in forecasted manhours versus the baseline.

A simple way of determining a well-defined and baselined scope and how behind or ahead of schedule the project is to draw a horizontal line between the most up-to-date earned progress, (The red line in Fig 11.9) and the baseline, (The black line in Fig 11.9) and then read duration using the chart time scale. Earned progress to the left indicates the project is behind schedule, and earned progress to the right indicates the project is ahead of schedule. This can be trended on a weekly basis to determine slippage or recovery trends.

Plan Risking

Both schedule and estimate risking are covered in the Chapter 14.

Planning Summary

Planning is a complex yet essential subject and worthy of further study and training by project engineers. Effective planning will:

- Coordinate the activities of your internal and external teams.
- Provide early warnings and allow for corrective action to be taken.
- Demonstrate to the client that you are in control of the project.

When you plan effectively you will sometimes give yourself a "fright" that something is running tight or worse late and requires corrective action. This is the goal of active planning and should be welcomed.

Fail to plan – plan to fail.

CHAPTER 12

Estimating

Estimates are essential in a project environment, both for the original base project scope and for any subsequent changes. They are the mechanism by which anticipated outturn costs are advised to clients and form part of the contractual agreement to proceed, expend resources, and commit costs to a project.

If your tender estimate is too high, you are more likely not to get the work. Too low and you may lose your company money. It is important to get as accurate an estimate as possible. This chapter will explain the process to prepare, submit and obtain an estimate approval.

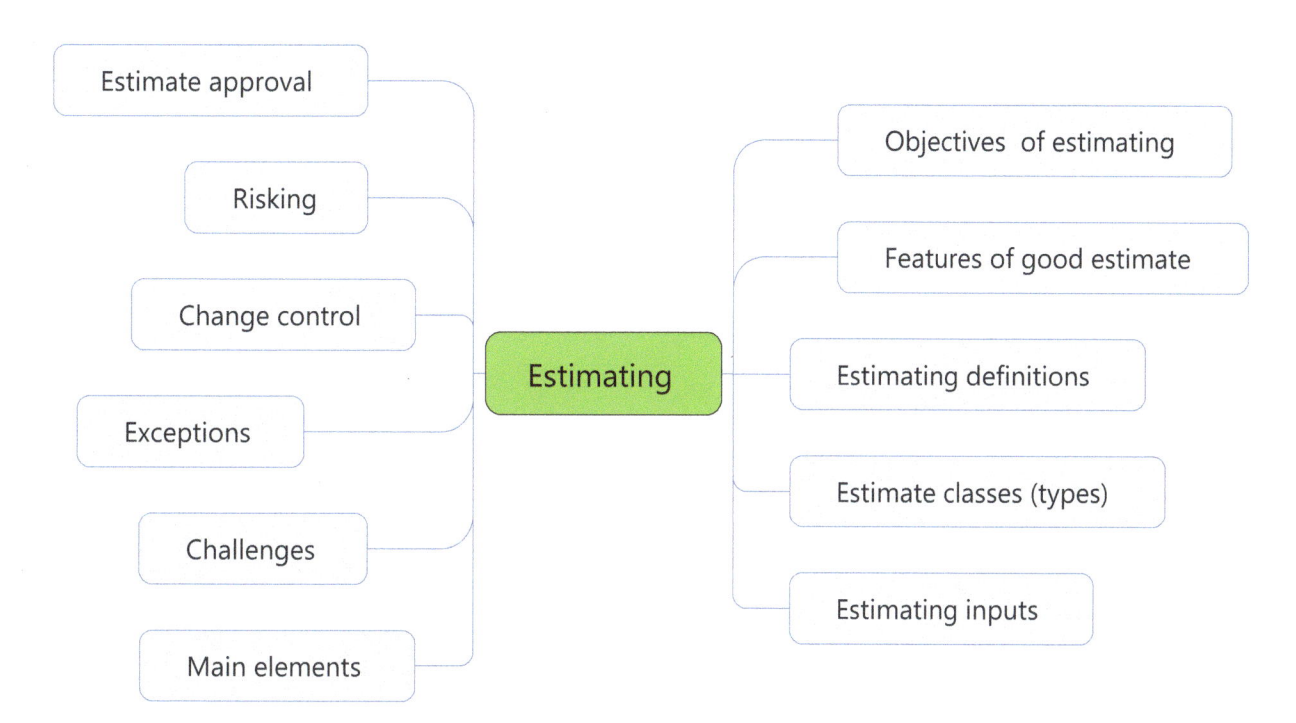

Figure 12.1 Estimating Mind Map.

CHAPTER 12 ESTIMATING

The Objectives of Estimating

The objectives of estimating are to:

- Provide input to the tendering process.
- Establish the capital requirement for the project phase or variation request.
- Provide a basis for change control.
- Generate a formal controlled estimate document.

The estimating team

Collating, generating, and approving the estimate is a team process. The Project Engineer will initially brief the whole team on the required scope and will collate all the inputs. When he/she is satisfied that the inputs are complete, he/she will issue the base figures to the estimator who will compile an estimate based on the organisations estimating procedure. In some organisations, Project Engineers will be responsible for compiling smaller estimates for minor scopes and changes, but normally an experienced estimator will compile the larger estimates.

Features of a Good Estimate

An estimate is only as good as the input data. The estimate will contain the following elements:

Direct costs: engineering manhours, materials/equipment, offshore construction manhours, etc.
Indirect costs: offshore supervision, temporary works offshore, offshore vendor support, etc. These are usually a percentage of the offshore construction manhours.
Overhead costs: office accommodation, IT services, corporate support, etc.
Below-the-line costs: These are costs that the client will pay for on the project that do not appear on the main contractors' invoices, such as offshore scaffolding and helicopter services which are normally billed directly to the client.

It may seem a little daunting when initially compiling an estimate, but most organisations have checklists which can be used to ensure that all potential areas of cost have been identified.

Tip: If your organisation or project does not have an estimating checklist, borrow one from someone who does and modify it to suit your organisation.

Estimating Definitions

There are many important estimating definitions you must understand. Unfortunately, all organisations are not consistent in definitions and use. For this overview, we will use the Project Management Institute (PMI), Book of Knowledge definitions(PMBOK). If you are in an organisation, I suggest you review your internal estimating procedure and adhere to their specific definitions.

Tip: Due to differing definitions regarding growth contingency and reserve, people from different organisations discussing an estimate may use the same words but have different understandings. When tabling your estimate, use a visual and be sure to explain your definitions.

The three main elements of an estimate are:

- Base project estimate that contains all the input activity costs.
- Contingency, which is the percentage added for all known risks.
- Management Reserve, which is the percentage added for unknown risks.

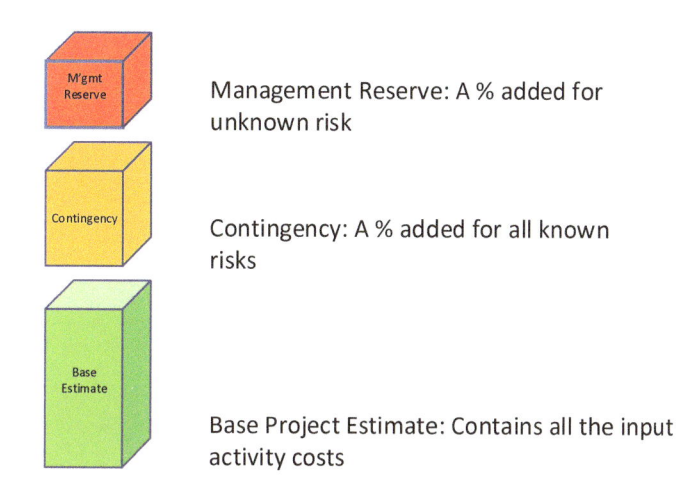

Figure 12.2 Main elements of an estimate.

Estimating definitions may change but the underlying principles are the same. The contractor Project Engineer will normally be tasked with managing the cost baseline and drawing down on contingency when known risks materialise. If an unknown risk happens, the contractor will raise an external change request to draw down from the client management reserve. Finally, the client project manager is tasked with managing the overall project within the cost budget.

Classes (Types) of Estimates

Estimates are required when evaluating concepts, completing studies, submitting a project for implementation approval, and for re-estimating an ongoing scope. It may take many months to complete a detailed estimate for a major project tender. Conversely, there may be a requirement to carry out comparative screening costs for 10 options and complete them in a matter of days. To allow fit-for-purpose estimates to be compiled and effectively used, a range of estimate classes has been produced.

CHAPTER 12 ESTIMATING

Estimate class	Name	Purpose	Project definition level
Class 5	Order of magnitude	Screening or feasibility	0% to 2%
Class 4	Intermediate	Concept study or feasibility	1% to 15%
Class 3	Preliminary	Budget, authorisation, or control	10% to 40%
Class 2	Substantive	Control or bid/tender	30% to 70%
Class 1	Definitive	Check estimate or bid/tender	50% to 100%

Table 12.3 Classes of estimate project definition

Estimate class	Name	Purpose	Expected contingency Range
Class 5	Order of magnitude	Screening or feasibility	<30%
Class 4	Intermediate	Concept study or feasibility	<25%
Class 3	Preliminary	Budget, authorisation, or control	10% to 20%
Class 2	Substantive	Control or bid/tender	10% to 15%
Class 1	Definitive	Check estimate or bid/tender	5% to 10%

Table 12.4 Classes of estimate expected contingency range

Features of estimates

- Concept/study type estimates can be completed relatively quickly and easily but with relatively low accuracy.
- The accuracy of an estimate increases as the definition improves.
- Base estimate cost normally increases as definition improves.

Estimating Inputs

The input data to a project varies depending on the maturity of project definition and the end usage for the estimate. The following techniques are used in preparation of different classes of estimate.

Parametric modelling

This uses statistical modelling to develop the estimate based on different parameters such as module dimensions, weight and equipment complexity.

Expert judgement

This uses the expertise of teams who have completed similar projects.

Analogous estimating

This uses historical data from similar projects, allowing for the estimate to be adjusted for known differences between projects. It is normally used in the early phases of a project.

Bottom-up estimation

This uses the estimates compiled for each WBS element in consideration of engineering hours, weights of materials, bought-in items, construction manhours, and so on. This is the most time-consuming estimate to generate but is also the most accurate and will normally be required to be completed and approved before a client releases funds to implement a project per the design/implementation phase.

Elements of an Estimate

All offshore engineering scope projects will have the following main elements:

- Engineering manhours.
- Materials and equipment.
- Subcontracts for onshore/offshore services and complex equipment.
- Offshore construction and commissioning manhours.
- Below the line costs including scaffolding, helicopters, and marine services.

Overheads

Some items such as project management, office services, and corporate departments are not estimated as line item costs but as overheads that are spread across all projects, typically resulting in a tariff/rate increase to the engineering hourly rates.

ESTIMATE COST ALLOCATION

Cost Elements	% Split	Total Cost
Engineering & Design	8.7%	£ 69,540
Follow On	0.4%	£ 3,148
Technical Close Out	0.7%	£ 5,494
Material	14.2%	£ 113,803
Fabrication	0.3%	£ 2,398
Subcontracts - offshore	8.7%	£ 70,040
Construction - Directs	5.8%	£ 46,151
Construction - Indirect & Non-Prods	8.1%	£ 64,611
Client Project Cost	27.8%	£ 223,272
3rd party - Scaffold / Fab Maint	2.6%	£ 20,674
Verification 3rd Party Costs	1.2%	£ 10,000
Contingency	21.6%	£ 172,998
Total	**100%**	**£ 802,387**

Figure 12.5 Typical summary cost estimate.

Engineering and construction norms

Norms are average estimates generated for all engineering and construction activities. For example, the generation of an engineering document may have a norm of 20 hours and the construction of 5 meters of 2" pipework may have a norm of 10 hours. These have been statistically developed over many years and are based on actual manhours from previous projects. Norms databases are maintained by the contractor and independent quantity surveyor organisations and are regularly reviewed and adjusted as technology and work practices evolve.

UNIT: No.   NOMINAL BORE

2nd REF	DESCRIPTION	3rd Ref INS	01 N.e 1.5"	02 2"	03 3"	04 4"	06 6"
00	SCH 5	No.	3.08	3.26	3.74	3.95	4.49
01	SCH 10	No.	3.08	3.26	3.74	3.95	4.49
02	SCH 20	No.	3.08	3.26	3.74	3.95	4.49
03	SCH 30	No.	3.08	3.26	3.74	3.95	4.49
04	SCH 40	No.	3.08	3.26	3.74	3.95	4.49
06	SCH 60	No.	3.08	3.26	3.74	3.95	4.49
08	SCH 80	No.	3.53	3.81	4.58	4.97	6.12
10	SCH 100	No.	3.53	3.81	4.58	4.97	6.12
12	SCH 120	No.	4.31	4.68	5.64	6.33	8.04
14	SCH 140	No.	4.31	4.68	5.64	6.33	8.04

Figure 12.6 Example construction norms for pipework.

Estimate Challenges

When you submit an estimate to the client for a complete project or a change approval, it is likely that it will be challenged by the client's representative. He/she will be tasked by his management to ensure a robust estimate, so it is incumbent upon the project engineer to ensure that the estimates presented have also been internally challenged and are as accurate as possible.

As you gain experience and get to know your teams, you will learn who underestimates, who pads estimates, and who gets it just right. The internal challenge should start once you receive the draft input from the technical disciplines.

How to challenge estimates

Review all the engineering hours and look at factored complexity, which is where the engineering disciplines can apply a multiple for perceived additional complexity. If the complexity is greater than one, challenge the discipline for a justification, and if they are unable to provide a solid reason that you can also explain to your client, ask them to reduce it. Repeat this process for the entire estimate.

Tip: Do not arbitrarily cut hours thinking it is good but tough management practice as it will result in an underestimate and likely cause an overspend, an unhappy client, and alienation of your engineering discipline team. Similarly, if you think something is underestimated or missed, discuss it with your team and to include or adjust it.

Strive to make the estimate as accurate as possible and be as firm but fair as possible when challenging your teams so that you can confidently stand in front of the client and address any subsequent challenges.

Exceptions

Now that you have completed your estimate, you need to be crystal clear about what is included and what not included in it. Reduce the grey areas to minimise arguments in managing subsequent additions to the scope and estimate.

Caveats

All estimates require the listing of all inclusions and exclusions to allow a complete picture to be made of the scope. This should be stated in detail as part of the submitted estimate to allow for future contractual reference. These are normally done per major element in the estimate.

Exclusions

Highlight the areas that you have identified that are specifically not included to ensure that an excluded area requires estimation and budgeting at a later stage in the project. This will be a formal record that no cost is included for these items

Provisional sums

Provisional sums are the "order of magnitude" or higher level sums for items required and budgeted for but are not sufficiently defined for lumpsum KPI agreements. The benefit to the client is that the sums can be included in their cost baseline to ensure that sufficient funds are approved by their management.

When you complete the caveats, there will always be some items you do not catch, however if done properly, you will minimise difficult conversations with your client.

Change Control

Now that you have completed a robust and accurate estimate and taken the time to generate all the estimate caveats, you will have a reference document to refer to for managing future change. When a change is identified or requested, determine whether it is in the base scope, and if not you may require a change/variation request. The change should then be estimated and submitted for approval, and should it be rejected, the change must not be implemented.

Estimate Risking

Estimate risking is covered in full detail in Chapter 14 and is similar in principle to schedule risking. Formal Monte Carlo risking of an estimate allows for contingency and reserve sums to be established by taking into account the contractor and client risk appetites.

Estimate Approval

When you have compiled your estimate and worked through internal and client challenges, the estimate will be formally submitted for approval. For large projects, the estimate and supporting documentation will be submitted to a contractor contract review board (CRB) that will confirm they are comfortable with the project estimate from a corporate compliance and governance standpoint. Once signed, it will be submitted to the client for signing, and the client will then authorise the contractor to commit expenditure up to the cost baseline.

Estimating Summary

Estimating is important. If you get it wrong you can cost your organisation money and impact the company's reputation, potentially even damaging your own career. The emphasis must be on a rigorous, structured approach that clearly states what is included and what is not, ensuring that the majority of your projects come within budget.

Project Within Budget? Yes. Happy Client! Happy Senior Management! And Happy Project Engineer!

CHAPTER 13

Cost Management

Businesses exist to make money that goes towards employee salaries, shareholder dividends, and reinvestment in the business. If costs are not controlled correctly, the business will be in trouble and potentially fail. This chapter will show you how costs are monitored and controlled, and how to use the information in the various cost reports that will be produced on your projects. A lot of project insights can be gained from a detailed understanding of your project cost report.

Figure 13.1 Cost management Mind Map.

The PE is responsible for not only the scope and schedule on the project but also the costs. A technically successful project is not actually successful if the budget has been blown!

Objectives of Cost Reporting

In simple terms the objectives of cost reporting are as follows:

To controls budgets

In order to control budgets, you must establish controls for each element of the workscope based on the client-approved target cost estimate. By measuring and monitoring costs you can take any required actions, maximising the probability of completing the project within budget.

To monitor and analyse costs

The three key items that will be monitored against the approved budget are:

- Forecasted costs.
- Expended costs.
- Committed costs.

This monitoring and analysis exercise is usually carried out by the project cost engineers who will identify any emerging trends and give early warnings of any issues.

To provide accurate and updated costs

Accurate costs are provided on a regular basis to allow clients to adjust their budget for reductions or add-ons. Underspending an approved budget can be just as bad as overspending as the client has lost the opportunity to invest the underspent budget.

To manage growth and offset allocations

Estimates will normally have a growth element which is a percentage of the budget uplift that is agreed on during project approval and can be allocated for any justified internal overspends. The cost engineer will keep an eye on growth and advise if it is being used up to early in the project. The cost engineer will also have discretion within agreed boundaries to offset costs, that is, to move money from an underspend on one part of the project to another part that may be overspending. This is known as an offset.

To provide regular updates to the contractor project team and client management

Normally this will take place monthly wherein the cost engineer produces the various costs reports for the complete project/projects and then meets with the individual Project Engineers and potentially Project Managers to review the draft report. This meeting will allow the project team to understand how they are performing against the approved budget and to understand any significant changes from the previous month.

To provide a variance narrative

One of the outputs from the cost meeting will be an agreed variance narrative used to brief the client about the cost report. A variance narrative is a written explanation of key changes and trends in the costs and the reasons for these changes.

To provide financial forecasts to corporate management

Provide information to corporate management on contributions to the overall income and the forecasts on cash flow.

Management adjustments

Management adjustments can be made to cost reports where a forecast cost increase is drip-fed into the forecast over a number of months, or an expenditure is re-phased based on management's judgement.

Tip: Being able to accurately explain all significant variances and their reasons, whether good and bad, to your clients will give them confidence that you are in control of the budget. The key to achieve this is to perform regular and detailed analysis and preparation.

Key Definitions

When I started my Projects career three decades ago, it took me quite some time to decipher the language of costs, and I recommend that you take time to understand the following key definitions that will occur on every project:

Cost control

The systematic restraint on expenditure to secure completion within pre-determined budgets or targets.

Approved budget

The funds approved and set aside as the authorised target for project achievement, usually expressed in terms of quantities, manhours, and costs.

Expenditure phasing

Time-based reporting usually shown as a graph that indicates when expenditure has been made.

Commitment

The total value of all orders and contracts placed to date, including letters of intent, but not including any allowances for anticipated additional costs in respect of such orders and contracts.

Value of Work Done

The value of work done (VOWD) is a project management technique for measuring and estimating the project cost at a given point in time. It is mainly used in project environments and is defined as the value of goods and services progressed, regardless of whether or not they have been paid for or received. The primary purpose of determining VOWD is to get an estimate of cost for the project that is as accurate and comprehensive as possible at a given point in time.

Anticipated Final Cost

Anticipated Final Cost (AFC), also called an Estimated Final Cost (EFC) is the forecast at a given point in time that takes into account expended costs, productivity, forecasted remaining costs, and any pending changes.

Cost Reporting Cycle

Costs reports for a project are normally produced monthly and will form a subset of the corporate group cost reports for senior management both within your organisation and the clients. There are many data streams inputted into a cost report that will have set cut-off dates in each month by which various elements must be completed, allowing all contributors to be coordinated.

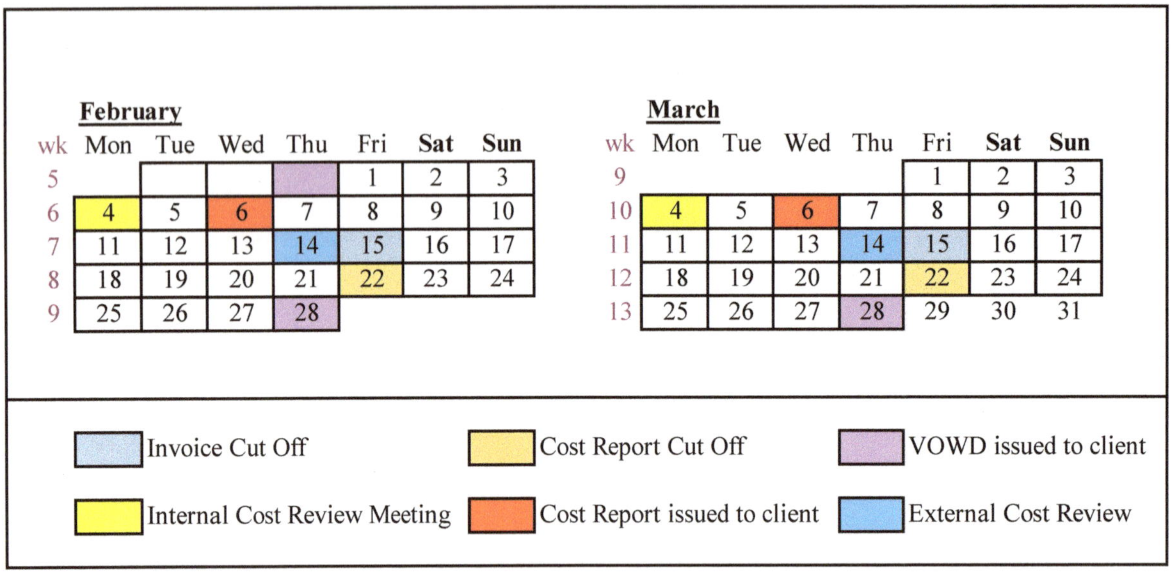

CHAPTER 13 COST MANAGEMENT

	February								March					
wk	Mon	Tue	Wed	Thu	Fri	Sat	Sun	wk	Mon	Tue	Wed	Thu	Fri	Sat	Sun
5					1	2	3	9					1	2	3
6	4	5	6	7	8	9	10	10	4	5	6	7	8	9	10
7	11	12	13	14	15	16	17	11	11	12	13	14	15	16	17
8	18	19	20	21	22	23	24	12	18	19	20	21	22	23	24
9	25	26	27	28				13	25	26	27	28	29	30	31

- Invoice Cut Off
- Cost Report Cut Off
- VOWD issued to client
- Internal Cost Review Meeting
- Cost Report issued to client
- External Cost Review

Figure 13.2 Cost reporting calendar.

Role of the Cost Engineer

The cost engineer maintains the cost reporting system for the project which may be on an excel spreadsheet, bespoke company software, or a commercial system. The cost engineer will normally organise and chair the regular cost meetings against the monthly cycle of cost reporting activities.

Normally all expenditure commitments will be checked and approved by the cost engineer prior to release. He/she will receive all requisitions and check that there is a budget for the item before making a commitment on the budget. The cost engineer will also be responsible for providing early warnings on any overspends and underspends.

Cost Management Module

A project cost management module will have numerous inputs and outputs form other project reporting and monitoring systems including:

- Onshore engineering timewriting.
- Materials and fabrication procurement.
- Subcontracts.
- Offshore timewriting.

A typical system is shown below. Note that these are company- and sometimes project- specific. The more automated the data transfer is between parts of this system, the more accurate and less time consuming the reporting will be.

THE PROJECT ENGINEER'S TOOLKIT•113

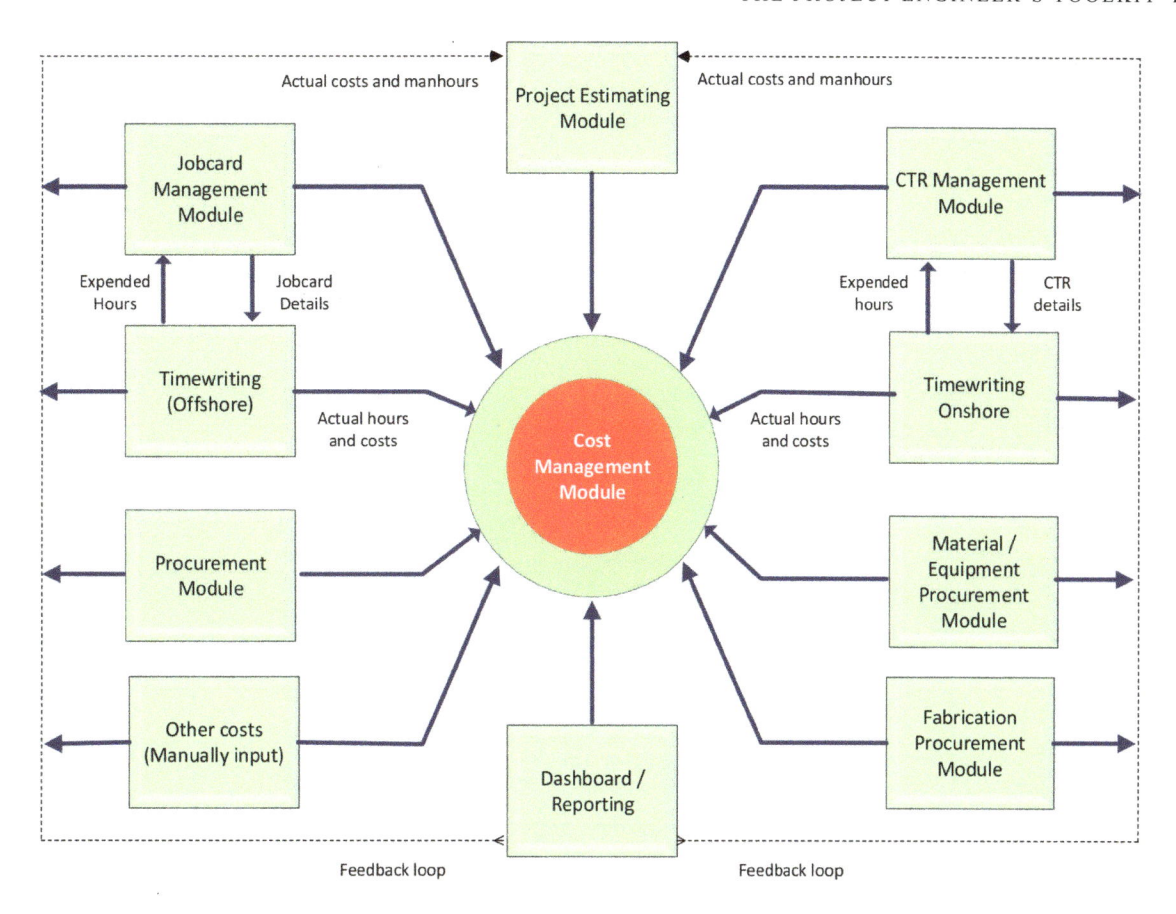

Figure 13.3 Cost management module

Using and Interpreting Cost Information

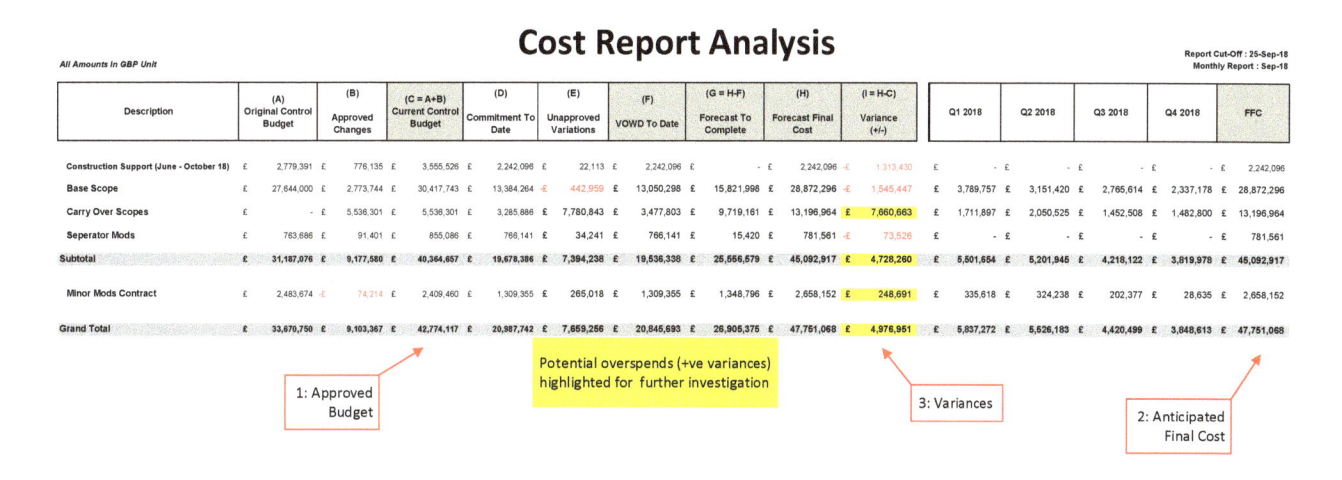

Figure 13.4 Simplified cost report.

Figures 13.3 and 13.4 illustrate the key elements of a cost report and how they are related.

CHAPTER 13 COST MANAGEMENT

Analysing a cost report

This is the easy way to quickly review a cost report and prepare for both discipline and client meetings:

1. Look at the bottom line and check against your approved budget.
2. Look at the AFC: If it is greater than the budget, you must understand why.
3. Look at the variance for the subtotals. A positive variance indicates a forecast overspend. Highlight these.
4. Look at line items for the highlighted overspend items from Step 3: Identify the offending overspend variances and highlight.
5. Make a list of the overspends. Your cost engineer might also do this for you. Discuss these items with your team and establish the reasons and required actions.

Forecast overspends

Overspend forecasts are often due to the following reasons:

- An underperforming discipline team or numerous changes in personnel within the team.
- Incorrect reporting of discipline progress by the lead engineer.
- Unapproved scope being worked on that requires a change request.

Cost curves

As in planning where reams of reports can be provided, simple graphics are available that allow easy and quick assimilation of the cost status of a project. The most commonly used is the cost S Curve. This a time phased plot showing as a minimum; the budget, commitments, and VOWD.

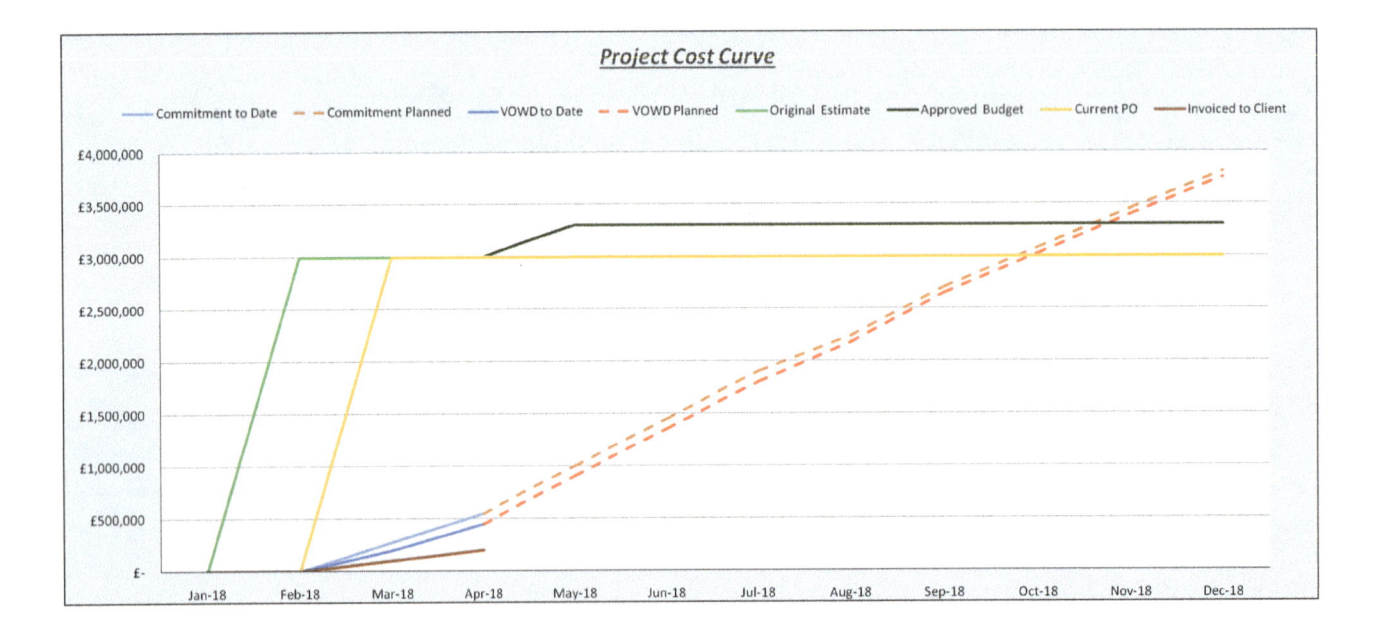

Figure 13.5 Cost curve.

PETER F CRANSTON

Simple rules for review of a cost curve

- Commitment and VOWD should always be under budget.
- Commitments should always be more than VOWD.

S Curves are often used in management briefings and can be provided for an overall project or split by project phase or engineering discipline.

Tip: When you attend the management briefing meetings, the Project Engineer will be expected to explain any issues or trends.

The S Curve can highlight issues such as:

- The project being behind schedule, shown by the amount of actual VOWD being less than planned VOWD.
- The project is overspending, shown by actual expenditure being more than planned expenditure.
- The project has unconstrained growth issues, shown by the slope of actual costs being greater than budgeted costs.

Cost Management Summary

With practice, reading and interpreting a cost report becomes easier, allowing you to ask the difficult questions. Think of the budget as your own money and how you would control and monitor it. The client is entrusting you to manage a significant portion, often many Millions of dollars, so please give it the attention it deserves.

Risk Management

All projects are risky as they are unique events subject to both internal and external events which are yet to occur. Risk management is an investment in your projects that both reduces probability and impact of risks and allows any opportunities to be maximised.

This chapter will explain the principles of risk management and allow you to immediately apply to your projects. I consider risk management as an invaluable alternative for reviewing your project because it identifies the required beneficial actions that may not be evident during the normal linear project management process.

Figure 14.1 Risk management Mind Map.

Definitions

Risk management

Risk management is a process that allows individual risk events and overall risk to be understood and managed proactively, optimising success by minimising threats and maximising opportunities.

Risk

Risk is defined as an uncertain event or condition that, will have a negative effect on one or more project objectives if it occurs.

Opportunity

Opportunity is defined as a risk that would have a positive effect on one or more project activities.

The Risk Management Process

In most organisations, the following general process is used to manage risk:

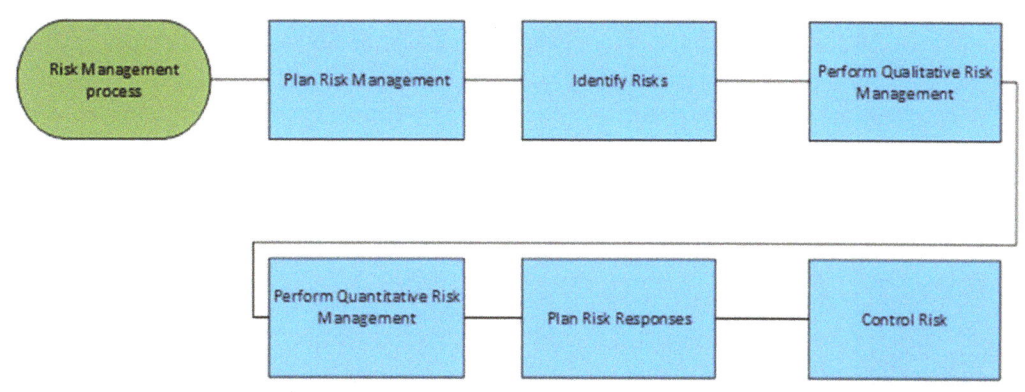

Figure 14.2 Risk management process.

Depending on the organisation there may be corporate risk management procedures and possibly project-specific procedures available that may be supported throughout the project by experienced internal or external risk facilitators. However, if the risk management documentation and dedicated risk facilitation personnel resources are not available, it is still beneficial for the PE to carry out the risk management him/herself.

Risk standard ISO 31000

ISO 31000 is a family of standards relating to risk management codified by the International Organization for Standardisation (ISO). The purpose of ISO 31000 is to provide principles and generic guidelines for risk management and can be used by any organisation, regardless of its size, activity, or sector.

Cost of risk management

Analysis has shown that correctly implemented risk management is an investment that will provide significant payback on both costs and schedule.

Qualitative Risk Analysis

Plan risk management

I suggest you generate a risk management plan outlining how you will manage risk. You may create it based on corporate guidelines, although should that not be available, you can follow the outline below:

Generate a short document detailing how you will carry out risk management:

- List of personnel to be involved in risk management.
- Roles and responsibilities for team members involved in risk management.
- Schedule for risk workshops and regular updates.
- A probability/impact assessment table (detailed later).
- Risk action tracking and reporting details.

Identify risks

Risks and opportunities are typically identified from individual or group review of existing available documentation and most frequently during group workshop brainstorm sessions.

The brainstorm is a particularly effective method for identifying risks as the team can "springboard" additional risks off each other. During initial brainstorming there should be no categorisation of probability, impact, or mitigation. The aim is to record as many risks and opportunities as possible.

A well-defined risk has the following 3 elements:

- A cause.
- An event.
- A consequence.

Probability/impact assessment table

The probability of an event occurring is given a score of 1 to 5 as shown below.

Probability		
Rating	Description	%
1	Improbable/Rare	<10
2	Unlikely	10-30
3	Possible	30-50
4	Likely	50-70
5	Almost Certain	>70

Table 14.2 Probability table.

The team is usually guided by either a risk facilitator or the project manager with regards to setting a scale or determining costs, time, and reputational impact of minor to catastrophic events, which is used in the ranking process. See Table 14.2 below:

Impact									
Opportunity					Threat				
Rating	Description	Cost (£)	Time Impact	Other Impact Area	Rating	Description	Cost (£)	Time Impact	Other Impact Area
-1	Minimal	<10K	1 – 3 days	TBA	1	Insignificant	<10K	1 – 3 days	TBA
-2	Minor	10K – 50K	3 – 7 days	TBA	2	Minor	10K – 50K	3 – 7 days	TBA
-3	Moderate	50K -100K	1 – 2 weeks	TBA	3	Moderate	50K -100K	1 – 2 weeks	TBA
-4	Major	100K – 250K	2 – 4 weeks	TBA	4	Major	100K – 250K	2 – 4 weeks	TBA
-5	Exceptional	>250K	>4 weeks	TBA	5	Critical	>250K	>4 weeks	TBA

Table 14.3 Impact assessment table.

Qualitative risk assessment

Once risks have been identified, the team works through each one to determine the probability of occurrence and the impact. Whilst this is not an exact science, the group will generally have a good idea of the risks location on the impact assessment table.

The resulting risk number is obtained by multiplying the probability (1 to 5) by impact (1 to 5).

These figures are then plotted on a heat map/Boston square, with the largest risks at the top right and lowest risks on the bottom left.

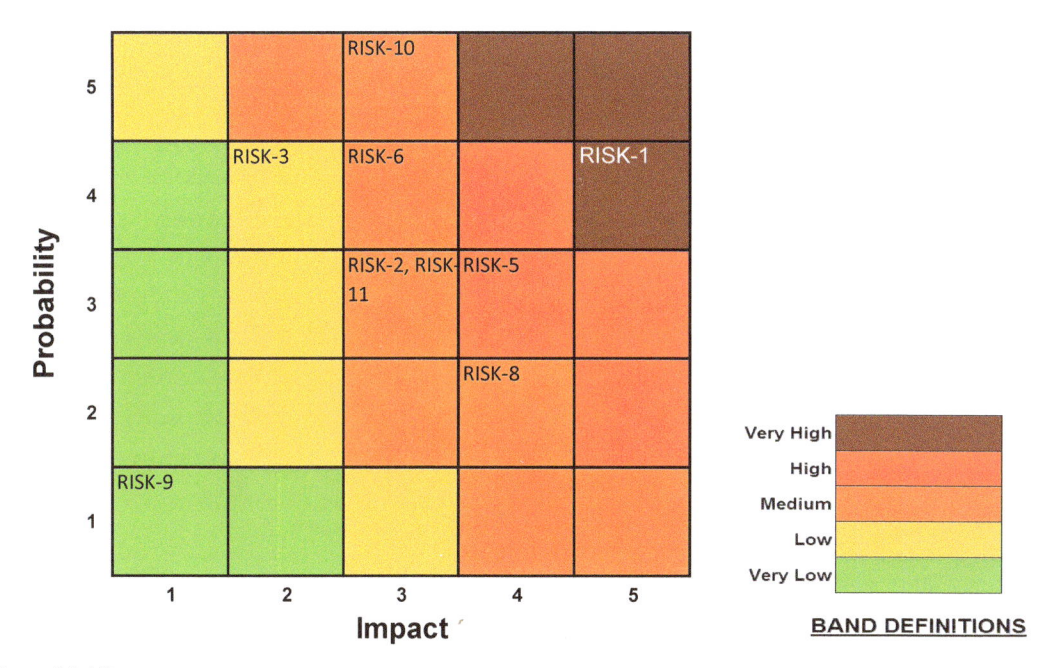

Figure 14.4 Boston square.

Risk responses

For risk management to be effective, we must do something with the identified risks. For each risk, assign an owner to develop a risk response, which should be specific and have a date for the actions/s to be completed.

Risk register

Risks should be documented within a risk register containing appropriate status information actions and the owner associated with the risk. The register may be a company-wide or project-specific register.

Risk mitigations

There are four main ways of dealing with a risk:

- Avoid.
- Mitigate.
- Transfer.
- Accept.

Avoid: This removes a risk by removing the work package, organisation, or person who is the source of the risk either by redesigning or by changing-out personnel.

Mitigate: This is where money or resources are invested to reduce the likelihood or impact of the risk.

Transfer: This is where the risk is transferred to another party.

Accept: Acknowledge the risk and accept the consequences if it happens.

Risk management examples

Mitigation example: To reduce the risk delay of equipment being shipped offshore on the regular weekly supply vessel, you might pre-invest in a dedicated standby vessel to transport the equipment when required.

Transfer of risk example: A warranty on an equipment package transfers risk to the supplier wherein the supplier repairs or replaces the product, part, or service should it fail during the warranty period.

Tip: If time/resources are an issue, I suggest you concentrate your efforts on the largest risks and opportunities, and review lower risks regularly to determine if their risk levels are increasing.

Contingencies

Risk contingencies are pre-agreed plans that will be implemented if a certain risk materialises. An example is the air compressor for a key construction operation of an offshore facility. A contingency plan may be to have a standby spare compressor for hire onshore and ready to be mobilised on short notice should the offshore unit fail.

Fall-backs

This is where you have a fall-back plan if the planned contingency does not work properly.

Tip: A significant amount of the success of a project depends on how much effort you have put into contingency and fall-back plans for your major risks.

Risk reporting

The risk register should be updated regularly, monthly at a minimum. Distribute the register to all action parties and stakeholders. I also recommend reporting the top 5 risks and opportunities to the organisation and client project management. The purpose of sharing the risks frequently is to:

- Provide clear visibility of key risks.
- Leverage support on mitigating key risks.

Now that we have seen how to complete qualitative risk assessments, we will move onto the quantitative risk assessments.

Quantitative Risk Assessment

Quantitative Risk Assessments are used to determine the probability of meeting project objectives, usually in terms of cost or schedule. We will follow the process for a cost estimate, but know that it is similar for schedule risking.

Deterministic estimate

The base "un-risked" cost estimate is compiled for a project by the project estimator and sometimes the Project Engineer. The total cost estimate will include all the individual estimates for each WBS element for engineering, materials, subcontracts, construction, etc.

Probabilistic estimate

Given that there will be known risks associated with the project, it is important to know how much contingency reserve to provide, which should be appropriate to the level of uncertainty in the project. Determine the contingency reserve by generating a probabilistic estimate that provides a range of outcomes with their associated probabilities, usually expressed as the P10, P50, and P90 values:

- P10: 10% probability of achieving this estimate.
- P50: 50% probability of achieving this estimate.
- P80: 80% probability of achieving this estimate.

Three-point estimates

The input to probabilistic analyses is a series of three-point estimates. The base estimate will be first conditioned to ensure that the quantity of input estimates is appropriately sized for both the risk workshop and the analysis software. We would not wish to risk 10,000 individual risk items, nor would twenty line items be sufficient for such a quantity. The base estimate is normally determined during a facilitated workshop with the relevant project team members supported by external expertise and review of each cost line item.

The team discusses each item and judges it based on experience and insight as to what the lowest cost for an element would be. They then repeat the process to determine the highest cost an element is likely to be. Repeat the procedure until all line items are completed. The output is three cost points for each line item, which is then entered into the Monto Carlo analysis software.

Monte Carlo simulation

Monte Carlo was first used by scientists working on the atomic bomb and is named after Monte Carlo, the resort town famed for its casinos. The Monte Carlo software performs risk analyses by building models of possible results by repeatedly substituting random values from the input three-point estimate. The result is a probability distribution.

CHAPTER 14 RISK MANAGEMENT

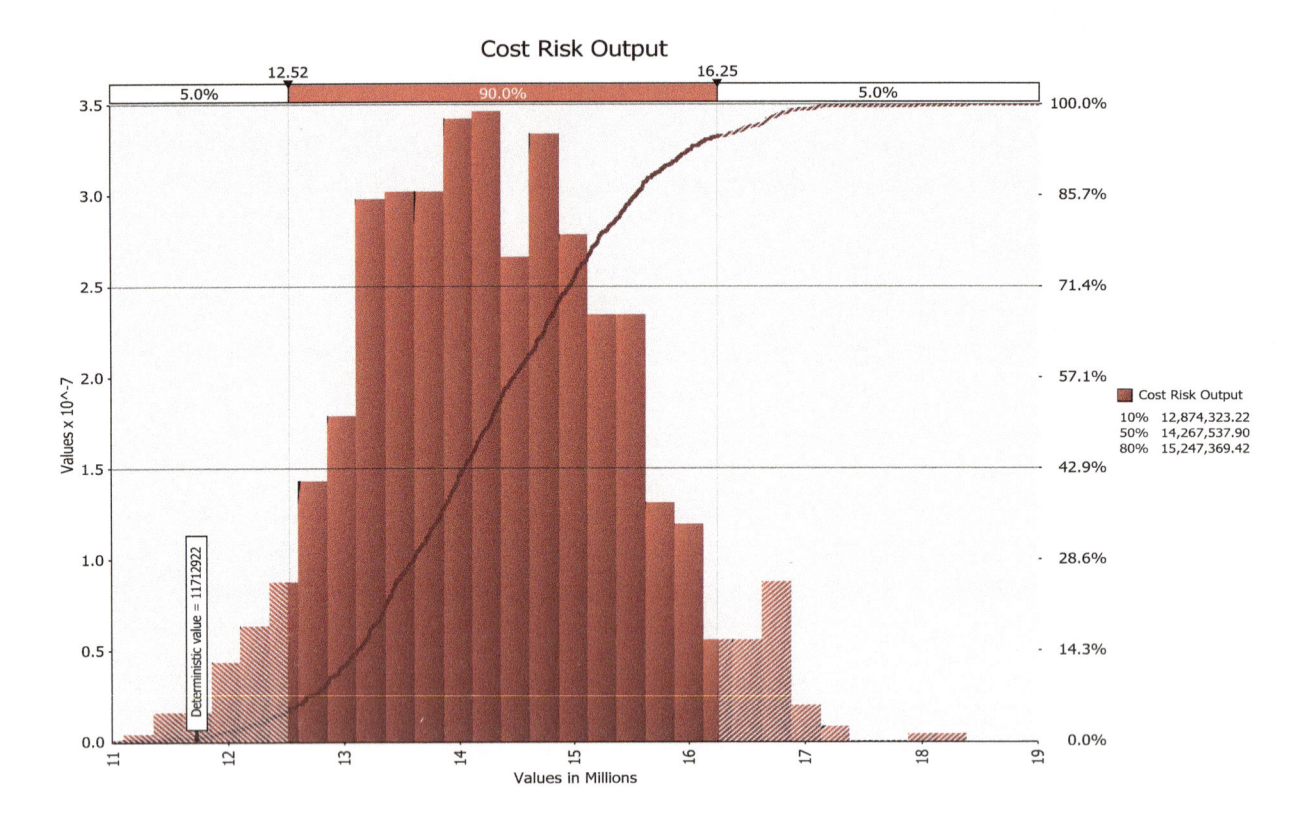

Figure 14.5 Monte Carlo probability distribution.

Consider a £10 million lump sum scope that you have estimated for. Would you be comfortable with submitting a P10 estimate (10% chance of coming in on budget)? Probably not. Similarly, you might be reluctant to go with the P90 estimate as this might be uncompetitive. Thus, the P80 (80% chance) might be considered the most realistic estimate to submit, especially if coupled with robust risk management. The final agreed-upon value will also be influenced by the organisation's and client's risk appetite.

A quantitative risk assessment for either a cost estimate or schedule estimate allows the contingency reserve to be set for known risks. Each organisation will have agreed risk thresholds that they have evaluated quantitatively are comfortable with before committing to an agreement with clients.

Insights into estimates and schedules

One of the useful outputs of a Monte Carlo simulation is a tornado diagram, which shows the largest potential impacts on cost or schedule. This analysis lets you focus on activity to mitigate the largest risk factors.

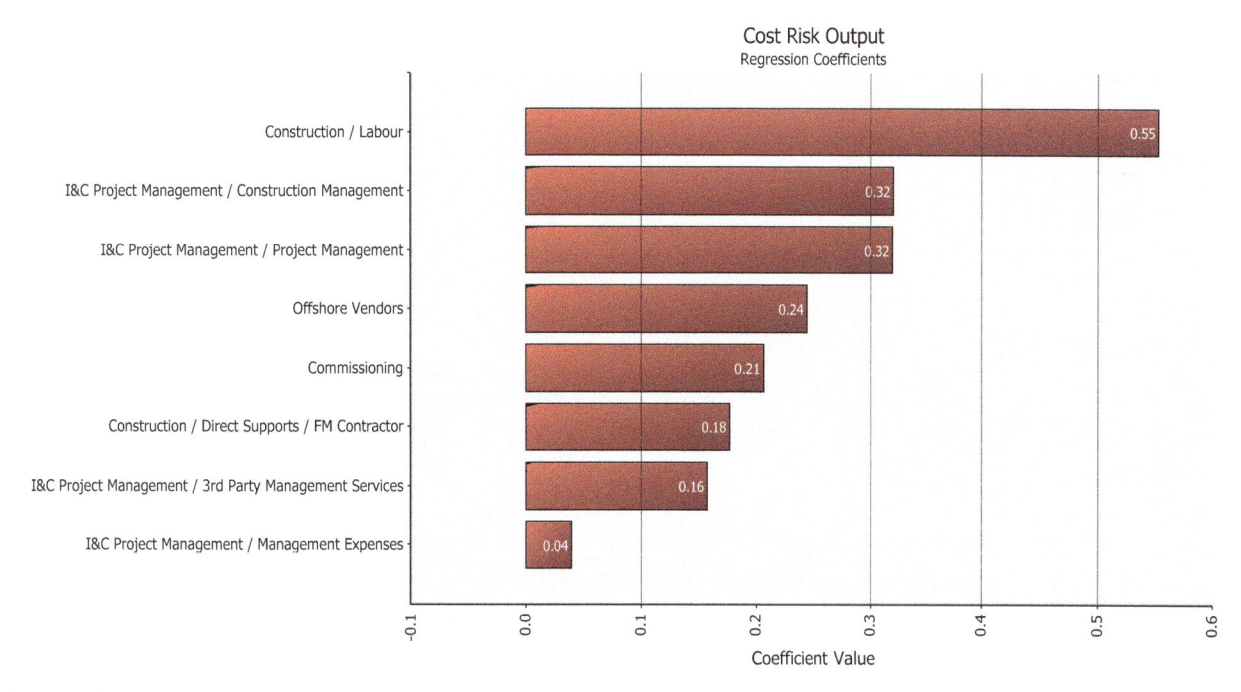

Cost Risk Output
Regression Coefficients

Factor	Coefficient Value
Construction / Labour	0.55
I&C Project Management / Construction Management	0.32
I&C Project Management / Project Management	0.32
Offshore Vendors	0.24
Commissioning	0.21
Construction / Direct Supports / FM Contractor	0.18
I&C Project Management / 3rd Party Management Services	0.16
I&C Project Management / Management Expenses	0.04

Figure 14.6 Tornado diagram.

Risk Management Summary

Risk management techniques are valuable on a project if used correctly to allow you to take cost-effective actions prior to an event happening. If a known risk materialises, you will already have an agreed method to move forward that can be implemented without putting the team under stress to instantly generate a solution. Risk management is both an art and a science. If you do not already practice risk management on your project, get started! You will be pleasantly surprised by the alternate perspectives it gives you on the problem areas and opportunities within your projects.

Keeping the risk management plan up-to-date can transform it from a door stop into a vital project management tool. Remember: what you don't know can kill your project.
- Bruce Pittman

CHAPTER 15

Supply Chain

The term Supply Chain, when applied to offshore topsides engineering, refers to a group of organisations, people, activities, information, and resources involved in moving a product or service from suppliers to the offshore customer, which is normally construction.

Supply chain activities typically involve:

- Procurement of materials, components, and equipment packages that are delivered to the end customer offshore.
- Provision of personnel and equipment to carry out specific physical activities offshore.
- Provision of engineering services onshore to carry out surveys, complete studies, and prepare reports.

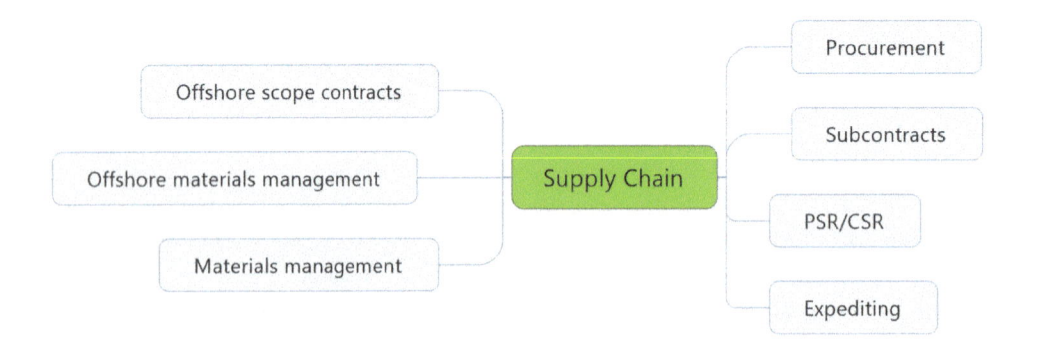

Figure 15.1 Supply chain Mind Map.

Procurement

The supply chain begins with procurement of either physical items or services. There are two main methods which will be explained in detail: Purchase Orders (POs) and Contracts.

The simple difference between purchase orders and contracts is that purchase orders are generally used for "off-the-shelf" procured items whereas contracts are used where there is either a bespoke engineering element of high value or where the terms and conditions associated with the procurement are complex.

In some organisations, POs and subcontracts are managed by separate teams, and for others the team working on the project manages them. However, it should be noted the negotiation and award of contracts will require the services of a contract engineer.

Both purchase orders and contracts have the following common features:

- A scope of work.
- A timescale for delivery.

We will now explain the processes associated with procurement purchase orders and contracts.

The Purchase Order

The purchase order is a procurement vehicle for typically standard, off-the-shelf items that do not have any complicated design and/or contractual requirements.

Purchase order process → Qualification → Enquiry → Technical and commercial bid evaluation → Terms and conditions alignment → Issue purchase order

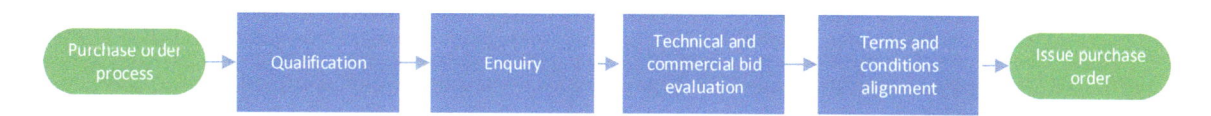

Figure 15.2 Purchase order process.

Qualification

Suppliers are normally required to be on an approved suppliers list, which is typically carried out by the company itself or the company's business unit. The procurement department, with help from the Quality Assurance (QA) department, will audit and select potential suppliers for quality, price, delivery, financial stability, etc. They then provide a list of pre-qualified vendors, which reduces the risk of future non-performance.

If a supplier is not listed as a preferred vendor, additional checks will have to be carried out, consisting of an audit of systems and review of publicly available company financial information. This pre-qualification of vendors ensures the risk of a non-performing vendor after an order is placed is reduced.

Enquiry

A requisition specifying the components/equipment required is prepared by the appropriate discipline engineer and will include a cover sheet, appropriate drawings, and specifications. The buyer will issue this out to a number of suppliers for quotation advising when they are required to respond with their proposal. More complicated scopes of supply typically allow weeks for suppliers to respond whilst "bulk" materials such as bolts and gaskets are given a few days.

Technical bid evaluation

The suppliers respond with their technical proposals and price to a technical bid evaluation. The technical proposal is issued by the buyer to the engineering discipline(s) for review, who will confirm whether the request is technically acceptable.

Commercial bid evaluation

Based on the suppliers that have provided technically compliant submissions, the buyer will then normally select the lowest-price and technically-compliant, proposal. The most frequent exception to this is when a shorter delivery is more important for a higher price to be justified.

Terms & conditions alignment

The terms and conditions associated with a purchase order are normally standardised for all items procured and are attached to the purchase order. These will contain key statements on payment and warranty. Preferred suppliers will normally have supplied the issued purchase order and be generally comfortable with the terms and conditions. Suppliers are required to comply with these terms and conditions and only in exceptional cases will these be varied.

Award

The PO is formally awarded by the buyer and the cost is committed to the cost management system.

In summary, a purchase order allows for a fairly quick enquiry and placement of the order, sometimes within a couple of days.

Subcontracts

Subcontracts are used where there is complexity and or high value in the scope of supply. The contracts process is similar to the PO process, although it is generally more prolonged and includes some extra steps.

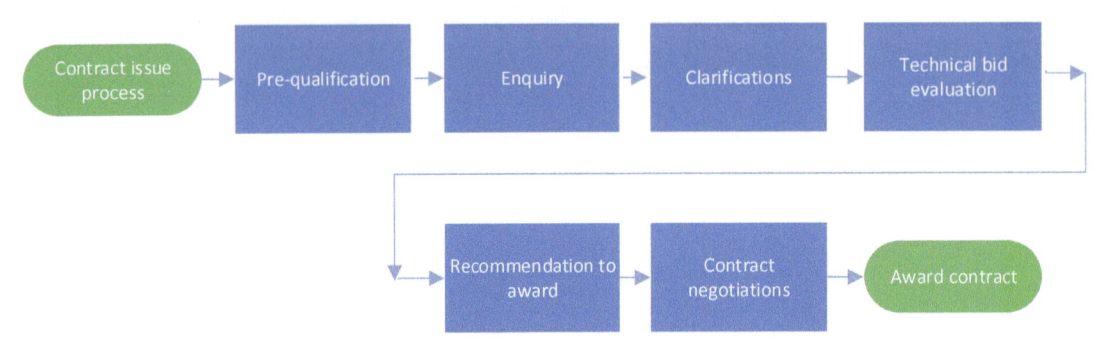

Figure 15.3 Contract award process.

Early in the project the Project Manager and the Contracts Engineer will prepare a contracting strategy which will determine the type of contracts that will be used for various elements of scope. This will include:

- Lump Sum Contract.
- Unit Price Contract.
- Cost Plus Contract.
- Incentive Contract.

The types and subtypes of contract and their applications are out of scope of this text, however, I suggest an internet search of engineering contracts which will provide additional information.

Pre-qualification

As tendering for major contracts is expensive for both the company and subcontractor, it is not beneficial to waste both organisations' time and money, and a fairly short pre-qualification questionnaire is issued to all potential tenderers and, based on their responses, a short list of "pre-qualified" subcontractors is produced to determine who will be issued the full tendering package.

Enquiry

The enquiry will be issued to all suppliers at the same time advising on date clarifications and acceptance deadlines for final submissions. It is not unusual for clarification periods and final tender submission dates to be extended due to the number of clarifications needed. A key feature of subcontract tendering is that all clarifications and responses are copied to all tenderers to ensure a level playing field.

Ethical behaviour

Given the often very large value of contracts, organisations and their employees must behave ethically in their relationships with suppliers. It is recommended that you complete your organisation's ethics training to determine what is and is not permitted in subcontractor relationships.

Unfortunately, there are some organisations who seek to influence decisions with gifts, money, etc and this is more prevalent in some countries than others. If in doubt, do not accept anything unless it is a small token item, approved by your manager, and declared in the organisation's conflicts of interest register.

Clarifications

Both technical and commercial evaluation normally require multiple rounds of clarifications, which are collated by supply chain and issued to all the tenderers for their information and response. As with purchase orders, no direct contact is allowed between engineers and tendering companies during the tender process to ensure improper influences or inducements are not made.

Technical bid evaluation

Similar to a PO, each tender is technically evaluated by the appropriate disciplines against pre-agreed criteria and a list of technically compliant tenderers advised to contracts.

Recommendation to award

In parallel with the technical bid, evaluation contracts will complete a pre-prepared commercial evaluation document which allows ranking of each tenderer. Normally, the recommendation to award will be made to the highest commercially ranking tender that is technically compliant. Contracts over a certain value require formal submission to a contract review board (CRB) comprised of senior management who will ensure that the proposal meets strict criteria and, when satisfied, will allow the contract to be awarded.

Contract negotiations

Unlike a PO where a standard set of T&Cs are used, a contract will require detailed agreement on all issues and items, such as payment schedules, reporting requirements, change management, warranties, and liabilities ,and in some cases may take many months for all contractual terms to be finalised.

Award

The final part of the process is the award of a contract, which is normally followed swiftly by a formal kick-off meeting with the engineering and commercial teams using a company standard agenda.

Features Common to both Purchase Orders and Contracts

Building in tendering time

It takes time to process requisitions, from preparation of the enquiry to placement of the order/contract, taking from weeks or months depending on the complexity and value.

Tip: Seek guidance from your supply team for appropriate lead time to account for in procured items.

Technical and commercial bid evaluation.

The normal process for receipt of tenders from suppliers is for the technical and commercial parts of the submission to be kept separate. Technical submissions are reviewed separately by the engineering team and are evaluated against pre-prepared ranking criteria. The engineer's job is to confirm which, if any, of the bids are technically compliant with the specifications.

Commercial information is withheld from the engineers to ensure they are not influenced by the process. Supply chain then selects a supplier from the list of technically acceptable submissions and awards based on price/delivery.

Single source justification.

The only exception to the competitive tendering process is when a single source justification (SSJ) is raised and approved. This is a document providing justification on why an order must go to a single company without following the normal tender process.

Justification may include:

- Established Contract or Agreement.
- Vendor has exclusive rights to supply/maintain.
- Required for technical compatibility with existing equipment.
- Research and development purposes only.
- Bargain purchase.
- Urgent requirement.

- An addition to an existing order Client requirement.
- Previously valid price/rate used.
- Known vendor with familiarity of client procedures.

Procurement Status Register and Contract Status register

These are the two most important supply chain documents for the project engineer. We will use the Procurement Status Register (PSR) as an example, but the same documents are used with contracts.

Description	Supplier		Enquiry Req Recd	Issue Enquiry	Receive Bids	TBE/CBA Issued	Req for Purchase	PO Issue	Lead-time (days)	ROS / Contractual / Forecast Date	PO Basis of Delivery	PO Float (days)
MOTORISED VALVES - 2 OFF	SEVERN	P	08-Aug-18	10-Aug-18	31-Aug-18	12-Oct-18	12-Oct-18	14-Oct-18	203	05-May-19	DDP	30
		F	23-Aug-18	23-Aug-18	18-Sep-18	15-Oct-18	15-Oct-18	19-Oct-18	168	05-Apr-19		
		A	23-Aug-18	23-Aug-18	18-Sep-18	15-Oct-18	15-Oct-18	19-Oct-18	168			
BURSTING DISC ASSEMBLY	ZOOK / BSB	P	07-Feb-19	08-Feb-19	13-Feb-19	18-Feb-19	18-Feb-19	19-Feb-19	70	30-Apr-19	Ex Works Forecast	8
		F	07-Dec-18	06-Dec-18	03-Jan-19	08-Feb-19	08-Feb-19	11-Feb-19	70	22-Apr-19		
		A	05-Dec-18	06-Dec-18	03-Jan-19							
INSTRUMENT FIELD CABLES	MACLEANS	P	21-Feb-19	22-Feb-19	27-Feb-19	04-Mar-19	04-Mar-19	05-Mar-19	56	30-Apr-19	DDP	-36
		F	25-Jan-19	22-Jan-19	30-Jan-19	05-Feb-19	04-Feb-19	05-Feb-19	70	05-Jun-19		
		A	21-Jan-19	22-Jan-19	30-Jan-19	04-Feb-19	04-Feb-19	05-Feb-19	70			
SAFETY SIGNS	TBA	P	21-Feb-19	22-Feb-19	27-Feb-19	04-Mar-19	04-Mar-19	05-Mar-19	56	30-Apr-19	Ex Works Forecast	20
		F	01-Feb-19	02-Feb-19	07-Feb-19	12-Feb-19	12-Feb-19	13-Feb-19	56	10-Apr-19		
		A										

| Not yet Started | Out for Enquiry | On Order | Delivered |

Figure 15.4 Procurement Status register excerpt.

The PSR lists all the purchase orders that are required to be placed for a project and is normally broken down by engineering discipline. It shows the planned, forecast, and actual dates for each step in the process. Most importantly, it shows the positive or negative float on delivery. A good PSR will be semi-automated, with macros raising red flags if deliveries are forecasting late.

Steps to building a PSR:

- At project kick-off, collate a list from the disciplines of all identified items that are required to be procured. This forms the basis of the PSR.
- Next, using the project schedule, determine when these are required on site (ROS).
- To determine the required delivery date, subtract at least 4 weeks for the ROS offshore date to give you float for any unforeseen events. Your supply chain will provide guidance on how much float you should build in for each type of procurement.
- You now have required delivery dates that can be communicated to suppliers.
- As orders are placed, you will get advised PO delivery dates which can be added to the PSR.

The most important activity is the weekly review of the PSR in which the PE and/or supply chain determine which items, if any, are forecasting later than the required delivery date. These are amber flags and should form a priority list for expediting and acceleration. Any items which are forecasting after the required on-site date should be flagged red for intervention to determine recovery actions.

The PSR management is an iterative process and should be carried out regularly, preferably weekly. You will develop a feel for where your procurement risks are and where additional effort or support should be directed. The key is to spend the most effort on those items that are forecasting late on their delivery dates and to keep a brief watch on the other items.

Expediting

When an order is placed, there will be a contractually agreed delivery date(s) for phased delivery scopes. We need to ensure that these orders and contracts are delivered on time to meet the project requirements.

Therefore, we bring in the services of the expediting team. There are three tiers to expediting:

- Report reviewing.
- Desk expediting.
- Field expediting.

Supplier Reports Review

When an order or contract is placed, there will be an agreed reporting format and frequency. This information is provided to the expeditor, the buyer and the procuring engineer.

Desk expediting

If there is any indication of slippage from the contractual delivery date, the expediter will contact the supplier by either email or phone.

Field expediting

In the case of the high value of highly critical items, the supply team will send in a field expediter to physically check the status of materials, fabrication, progress, etc.

By applying this phased approach, the project engineer can get early warnings of any issues that might affect the project. Good expediters are worth their weight in gold as they "cajole" suppliers into delivering in line with the agreed schedule.

Onshore Materials Management

Materials management is a sub-department of the supply chain that takes ownership of materials that are delivered and looking after them until they are delivered offshore to the construction team. The onshore materials management team are advised of final inspection and goods-release by the buyer.

Materials will not normally be released until all the agreed documentation (material certificates, inspection, and test certificates) have been reviewed and accepted. Materials are then physically shipped either to the companies directly or the client's warehouse. Note that some engineering contractors have their own warehouses and some share clients.

Goods inwards

Upon arrival at the warehouse, goods will be physically checked and logged into the usually computerised materials management system.

Preservation

Some components and equipment packages may not be intended to be used for many months and will require preservation to ensure they are fit for service when finally installed. Preservation may include filling with inert gas, silica gel, or preservation oil, or require regular rotation of shafts. An alternative to preservation in the warehouse is to keep these items on the supplier's premises for them to be responsible for until call off when required.

Warehousing

Warehouses are normally laid out with separate areas that are subdivided into shelves or bays for each of the intended offshore installations projects.

As you might expect, the majority of the warehousing process is computerised for easy tracking and shipment of materials, and includes increasing use of barcodes and radio frequency identification devices (RFIDs) to monitor the location and transfer of materials.

Bagging and tagging

As the name suggests, this is where components, usually smaller items such as nuts and bolts, and gaskets, are physically bagged with an identification of the workpack and jobcard that they are intended for to ensure a 100% pre-shipment check is carried out and any shortages are identified before time and cost-critical offshore operations occur, such as shutdowns.

Free issue items

A free issue item is one that has been procured by the client without an engineering contract and is "free issued" to the engineering contractor to install or use offshore. The project plan should track free issue items to ensure that they are visible and can be shipped when required.

Containerisation

Materials and equipment are normally shipped in robust steel containers designed for marine transport and offshore lifting operations. These are either fully enclosed or open "half height" containers for non-weather sensitive items.

Manifesting

The final activity by the onshore materials team is preparation of a manifest of all items within a particular shipment that will be passed to the marine movements team, who will in turn allocate space on a scheduled supply vessel. Manifesting is simply a list of materials in a standard format that suits shipping logistics, customs & excise, etc.

Offshore Materials Management

Supply vessels

All installations have regular supply vessel shipments, the frequency of which depends on how much activity there is onboard. Supply vessels position themselves close to the platform and keep position dynamically by use of their thruster systems. Weather is a major factor in offloading offshore; if the wind strength or sea-state are above prescribed limits, lifting operations cannot take place and the supply vessel must either wait on location or return to harbour.

Deck space management

Containers are landed on the deck of the facility using the offshore installations cranes. The location and movement of containers offshore can be likened a game of "three dimensional chess" with never enough space. The most common solution is to land a container and decant the contents into stores or directly at the worksite, and then back load the container as quickly as possible.

There are times when materials and equipment are shipped offshore for planned operations or construction, but due to other priorities there is not sufficient space, so they are then returned onshore. This is known as "round tripping".

Offloading limitations

Weather affects heavier loads because they must be lifted using the twin or triple fall hoist on the crane, which lifts more slowly due to the heavier load capacity, resulting in greater impact from wave motion.

Material storage

After offloading onto the offshore installation, the materials are either located directly to the worksite or temporarily located in stores until required.

Backloaded material

Waste and destructed materials produced during construction are normally segregated into appropriate containers and backloaded for disposal onshore.

Hire equipment

A significant amount of temporary equipment, pumps, compressors, lighting and so on, are hired and can be located offshore for many months.

Tip: It is important to keep track of hired equipment and off-hire it as soon as possible to avoid continuing hire charges. Hire companies want to make a profit, and it is not in their interest to remind you that you still have a piece of equipment.

Material shortages

Critical materials shortages can be procured on an urgent basis and prioritised for shipping. For really urgent, relatively light items, helicopter air freight can be requested, and normally needs the Offshore Installation Managers approval.

Contacts in place for offshore activities

When any subcontractor is mobilised offshore, the project must ensure that there is **always** a valid contract in place for that service. Without it, the engineering contractor and operator can be liable for any claims should damage or injury occur involving the subcontractor. Due to haste, many project engineers may not ensure that the contract is in place before mobilisation, thus exposing their organisation to claims.

Summary

The supply chain associated with offshore projects is complex and involves many organisations and people. For projects to remain on track, your materials and services must be delivered offshore to meet the required schedule. The key to this is an adequately resourced procurement team and weekly systematically review your procurement status register (PSR) to identify items that require additional focus or expediting.

Chapter 16

Change Control

You have worked hard to scope, estimate, and plan your project. You have finally obtained approval from both the client and your own management, and you have conducted your kick-off meeting. The next stage is to dust down your most powerful tool in the "Toolkit": Change Control. If you do not currently have a formal change process, do not worry as the principles outlined here will allow you to create your own simple process and procedure.

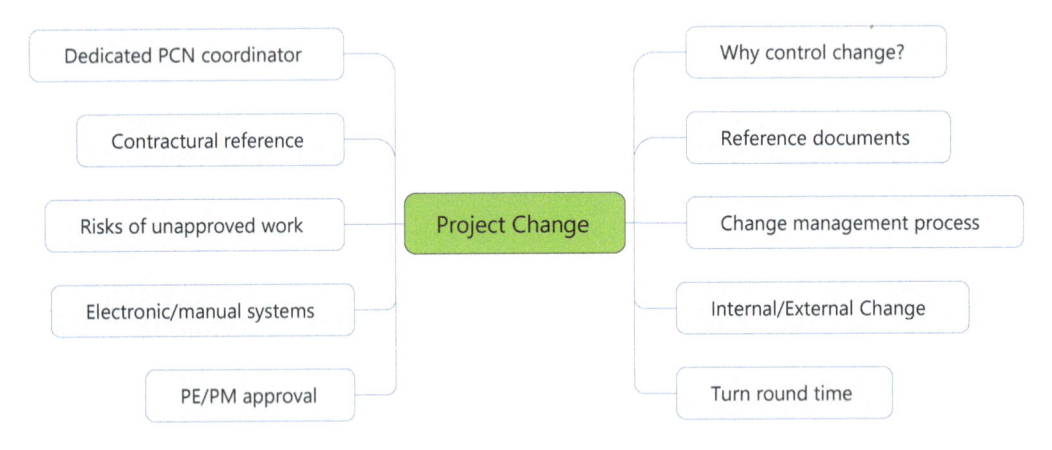

Figure 16.1 Change control Mind Map.

Why do We Need Change Control

Change happens in projects; it is inevitable because:

- Your team may have missed scope.
- The client might request extra scope to be added.
- The design development might dictate a change is required.

Almost all projects will have cost and schedule targets that may be fixed as part of a lump sum scope or linked to Key Performance Indicators (KPIs). Without formal change control, you are at risk of losing your organisation both money and reputation. Good change control will eliminate, or at least minimise, unapproved work. Change control will also be linked to both cost control and schedule control systems.

Change Control Reference Documents

To be able to identify a change in a project, you need to reference it against something because it is useless to argue with a client about a change unless you can clearly show via documentation why there is a change.

The following documents allow you to reference what is in the scope and what has been specifically excluded from the scope. This is where it is important to produce accurate and quality documents when completing scope and Front End Engineering Design, FEED, documentation.

Client documents

- Scope of work.

Contractor documents

- FEED report and attachments.
- Estimate.
- Plan.
- Decision register.
- Minutes of meetings.
- Emails.

Tip: Take care to fully document meetings in which decisions impacting scope, cost, or schedule have been verbally agreed upon as some clients can have convenient "professional memory lapses" when requested to retrospectively authorise additional costs committed in good faith. You will get no thanks from your own management either for spending money you do not have.

The Change Process

The change control process will require you to prepare change requests, also known as variation requests. Most organisations procedures follow this simple flowchart:

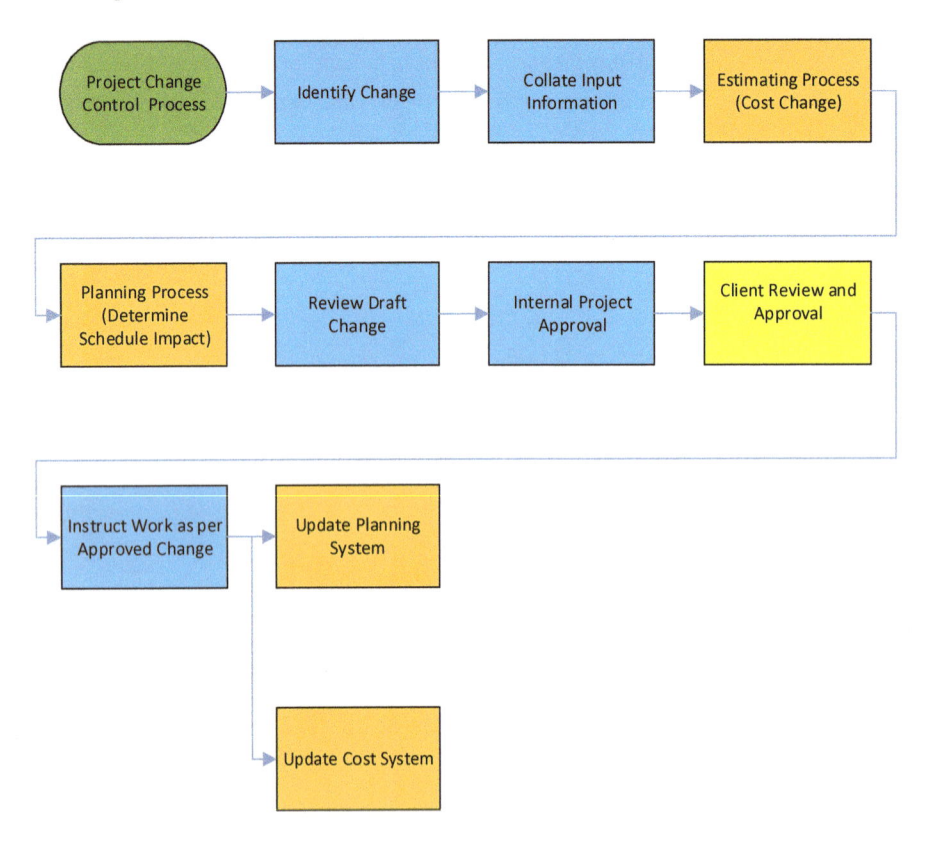

Figure 16.2 Change control flowchart.

Collate information regarding a change

This includes a clear scope statement, engineering man-hours, materials estimates, subcontract costs construction man-hours, etc.

Implement cost sub-process

Most organisations will have a cost-estimating procedure for changes. Think of a change as a mini–project containing the elements of scope, cost, and schedule. Your estimator will prepare a change estimate for the change and submit for your detailed review.

Implement planning sub-process

The change will run through the planning process to evaluate the impact on schedule dates and resource requirements.

Submit change for approval

Tip: As the PE, you must review the change and be confident enough in it to be able to stand up in front of your Project Manager and/or client to justify the change. Both Project Managers and clients can be very challenging when money is being requested or a schedule change proposed.

Approval/rejection

The change will be either approved or rejected. If approved, you may then update the cost and planning systems and instruct the additional work to be done. When change requests are rejected, it is often with comments on areas of challenge. You will then need to rework and resubmit. Do not commence work without an approved change or your Project Manager's written authorisation as you will be spending money that you do not have and may never get.

Types of Change

There are two main types of change: internal and external changes.

Internal change

This is required to complete the project as intended and might have been unintentionally omitted from the engineering scope, for example, the structural engineers have forgotten to include the pipe supports for a piping system.

External change

This is an additional request from the client that was not initially in the base scope, for example, the client now wants 3 gas turbine systems rather than the original 2 gas turbines.

Internal changes are funded from the project contingency and external changes from the client's management reserve. Unfortunately, change types are often not black and white as in the examples provided above.

Tip: Do not get into heated arguments about changes but instead refer to the scope documentation listed earlier.

Change Request Turn Round Time

There are many projects with change requests that take months to process and millions of dollars of unapproved work to implement. A good project will aim for change request turnarounds in circa two weeks from identification to approval, and will track the value of unapproved work that is being implemented. A good "yardstick" on how well a project is being managed can be gained from review of the change control reports and statistics.

Project Engineer & Project Manager Approval

The normal process is for the Project Engineer to be both knowledgeable about the change and involved in preparing and challenging the input. He/she must be able to discuss and explain any aspect of the change supported by appropriate disciplines.

CHAPTER 16 CHANGE CONTROL

As the PE, if you are happy with the change request documentation, you will endorse it. Before submitting the change request to the client, your Project Manager must review and approve it from the contractor's point of view. He/she will have an overview of the complete project and an understanding of any commercial or political sensitivities that should be taken into consideration. They may request re-wording or removing certain costs at their discretion, although they are relying on you for the detailed understanding of the change.

Manual and Electronic Change Control Systems

I have worked with both manual and electronic systems, and each has its pros and cons. A manual system is flexible and can be implemented almost immediately with minimal training and support while electronic systems are great for distributed teams, but you must have a tried and tested process to prevent glitches and maintain the efficient flow of information.

Tip: As a minimum, I recommend that you have the following:

- Change register for the project.
- Electronic folder for each change.

The change register

The change register is where each change is given an incremental number and has the following key information:

- Person/organisation requesting change.
- Title and short description of change.
- Internal or external change.
- Cost impact.
- Schedule impact.

PROJECT CHANGE REGISTER

WP Unit No.	Asset	Categorisation	Stage/Phase	PCN No.	Int / Ext	Description	Status	Overall Cost Impact
C009	Chieftan	MOD	DD	N/A	Ext	Gas Import Line	App	£ 715,796
C009	Chieftan	MOD	DD	001	Ext	Gas Import Line - Drain Line Clash	App	£ 51,673
C009	Chieftan	MOD	DD	002	Ext	Gas Import Line - Re-Route	App	£ 51,846
C009	Chieftan	MOD	DD	003	Ext	Gas Import Line - Phased Array Testing - approved by client, awaiting PO	Client	£ 13,151
C009	Chieftan	MOD	DD	004	Ext	Gas Import Line - Fire and Gas	PE	£ 91,043
C009	Chieftan	MOD	DD	005	Ext	Gas Import Line - Metallurgy & Piping Hours	PE	£ -

Legend:

	Estimate with client for approval
	Estimate has been approved
	Details in tracker for Info
	Estimate with Project Engineer for review

Figure 16.3 Change register sample

The change register can be in the form of a spreadsheet, database, or proprietary system.

An individual electronic folder for each change

Having an individual electronic folder documenting each change allows you to collate all e-mails, scope documents, cost information, and other documents relating to the particular change, providing a solid audit trail of information relating to the change.

The Risks of Unapproved Work

There are three main risks of unapproved work:

- Unpaid invoices.
- Claims.
- Reputation damage.

Unpaid Invoices: Unpaid invoices will impact the organisation's cash flow where the contractor has paid for services and has not yet been reimbursed.

Claims: Aggressive clients will reject late changes and counter the claims. If you have completed the work already, you are in a very weak negotiation position.

Reputation Damage: Significant unapproved work will indicate that the project is not being well-managed, thus damaging the Project Management's team's reputation as well as the company as a whole.

The Audit Trail

If the client brings in a firm of quantity surveyors at the end of a significant project with the objective of recovering cost via claims, it will be much harder for them to recover costs if you have a solid, fully-documented, and up-to-date change control system. Unfortunately, it does not matter who is in the right or wrong, but who has the best audit trail.

The Project Change Coordinator

On many projects it will be the project engineer who is responsible for maintaining the change management system, although you may be fortunate and have a dedicated change control coordinator or engineer. This is actually a great position for a graduate or entry level Project Engineer as they will get:

- Exposure to detailed scopes across the project.
- Exposure to the complete engineering and support teams.
- Exposure to the client.

Change Control Summary

Done well, change control can help you bring the project in on-time and on-budget. Done badly and you will lose the organisation money and impact its reputation negatively. A solid audit trail for each and every change will make commercial and contractual discussions at any point easier and much less stressful. Ultimately, your client will thank you for well-managed change.

Regarding change:

"The path of least resistance is the path of the loser." - H. G. Wells

I take from this to do the difficult and unpleasant change management tasks first as they will have the most benefit to the project.

CHAPTER 17

Project Management Organisations and Bodies of Knowledge

Compared to other disciplines, Project Management is a fairly young speciality. The USA-based Project Management Institute (PMI; 500,000 members worldwide) was founded in 1969, and the UK-based Association for Project Management (APM; 22,000 members, principally-UK based) in 1972.

Figure 17.1 PMO Mind Map.

These organisations have now grown and matured and through the expertise of their members, have developed recognised qualifications, information resources, and "bodies of knowledge". There are over 70 national project management organisations worldwide, each of which are members of the International Project Management Association (IPMA).

Bodies of Knowledge

"Bodies of Knowledge" are reference documents based on contributions from many project management specialists and can also be used as frameworks with their application tailored to a specific industry, company, or project. The frameworks provide guidelines for managing projects and defining project management concepts and processes, and are packed with references for further reading. They also provide a common language for projects with specific definitions to create a standard for all to understand rather than different individuals using their own descriptors that cause communication difficulties.

The two most prominent organisations in the UK, the APM and PMI, did not publish their composite bodies of knowledge until the 1990s. The APM body of knowledge was first published in 1992, and the PMI Body of Knowledge in 1996. Both organisations have updated and revised their bodies of knowledge over the years and most recently have added sections on the application of Agile techniques to project management.

CHAPTER 17 PROJECT MANAGEMENT ORGANISATIONS AND BODIES OF KNOWLEDGE

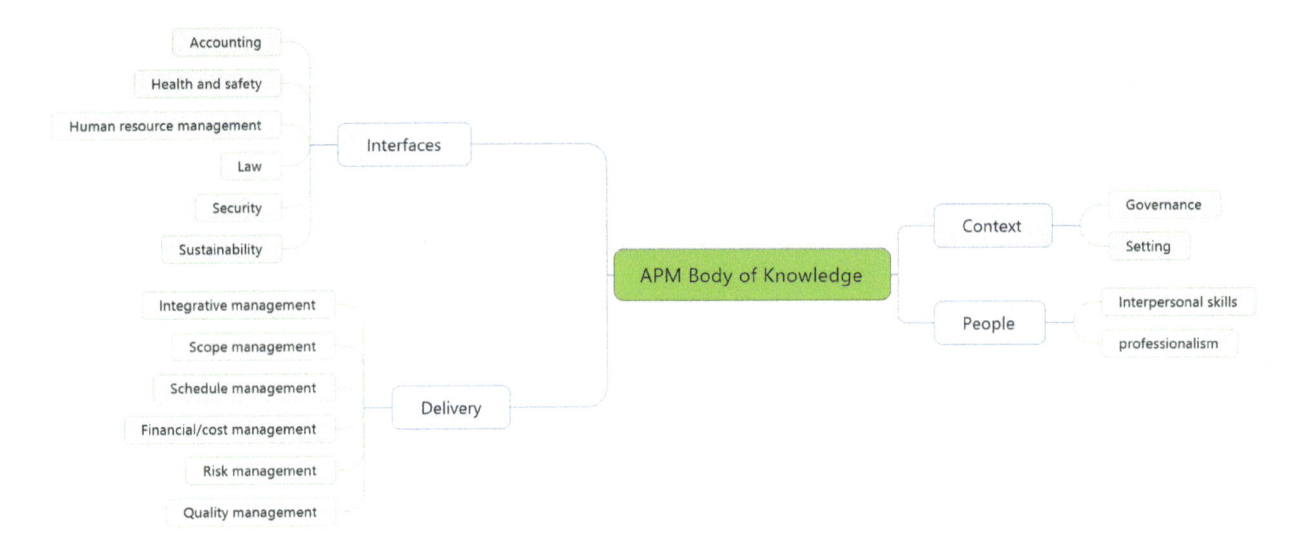

Figure 17.2 APM body of knowledge contents.

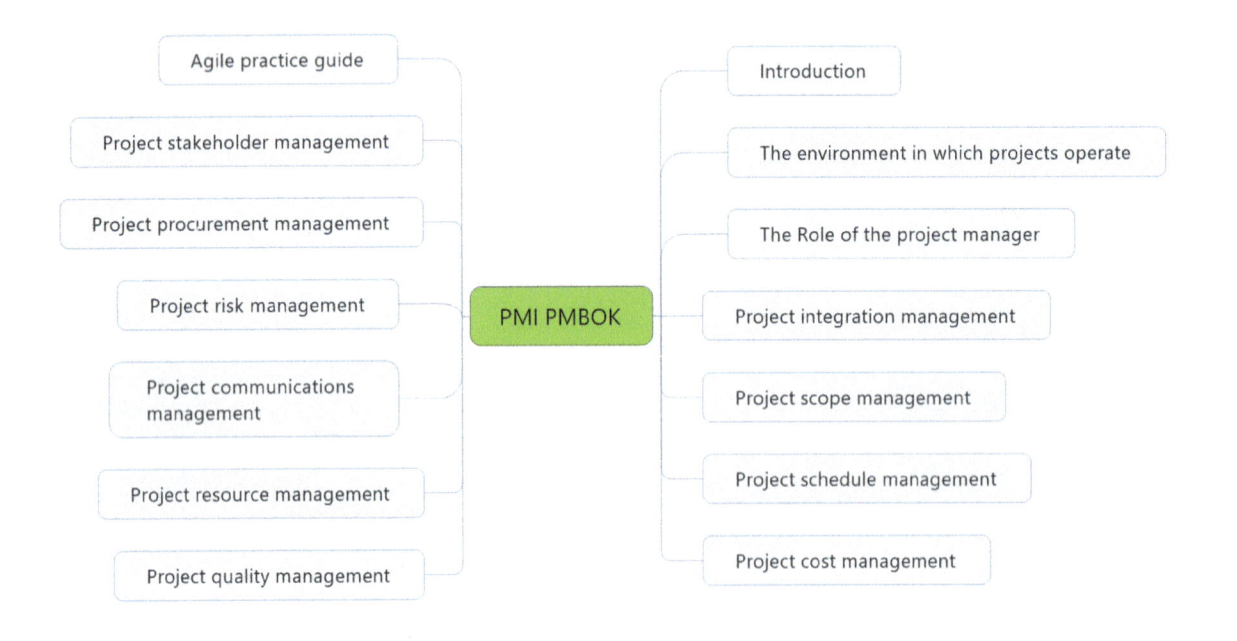

Figure 17.3 PMI body of knowledge contents.

Benefits of Membership

For the practicing Project Engineer, there are many benefits in becoming a member to either or both project management organisations, including access to:

- Written papers on specialist areas.
- Webinars covering a wide range of subjects.

- Project Management conferences.
- Latest profession developments i.e. "agile" and "lean" techniques.
- Networking opportunities and potential career opportunities.
- Cross-industry knowledge share.
- Volunteering opportunities.

Which Organisation to Join?

This will be mainly influenced by your employing organisation:

- Will your organisation pay for membership?
- Is there a local chapter/branch that is active?
- Is there a requirement to have a particular qualification to meet the job criteria?

What they all have in common is a structured qualification system, bodies of knowledge, and standardised project management vocabularies. Note that membership fees are normally tax deductible.

Tip: Join one project management organisation and attend as many local events as possible to widen your appreciation of project management in areas other than your speciality.

Project Management Qualifications

There are many excellent project managers who have no formal project management qualifications. However, it is increasingly being required to demonstrate competency via certification and this trend is likely to continue as project management matures further.

Master of Science in Project Management

There are many good Master of Science (MSc) courses in project management that take typically one year to complete if attending full-time, and three years to complete if attending part-time. For working Project Engineers with work and family commitments, the requirements and length of time can be a stressful route for qualification. I really admire all those who work full-time as Project Engineers while studying for qualifications part-time.

Some people take an MSc in project management immediately after their first degree without any significant project experience. I personally think that to maximise benefit from an MSc, you should work for a number of years in an engineering or projects environment so that you can personally relate to what is being taught while also developing the important people and life experience skills that are a significant part of Project Engineering.

Project management organisation qualifications

All project management organisations have structured qualifications based on a combination of theoretical knowledge, experience, and examinations that are graded from entry-level with little experience to highly experienced. These qualifications are not specific to the topsides oil and gas industry and are transferable across industries, which is good to keep in mind when there is an oil and gas downturn!

Association for Project Management qualifications

- Project Management Fundamentals (PFQ).
- Project Management Qualification (PMQ).
- Project Professional Qualification (PPQ).
- Practitioner Qualification (PQ).
- Chartered Project Professional (ChPP).

Project Management Institute qualifications

- Certified Associate in Project Management (CAPM).
- Project Management Professional (PMP).
- Program Management Professional (PgMP).
- PMI Agile Certified Practitioner (PMI-ACP).
- PMI Risk Management Professional (PMI-RMP).
- PMI Scheduling Professional (PMI-SP).
- Portfolio Management Professional (PfMP).
- PMI Professional in Business Analysis (PMI-PBA).

International Project management Association qualifications

- Certified Projects Director (Level A)
- Certified Senior Project Manager (Level B)
- Certified Project Manager (Level C)
- Certified Project Management Associate (Level D)

Each of the above has different education, experience, and continuing professional development requirements. Whilst these are useful in demonstrating a certain level of competency, you will need to back up your level with the ability to deliver projects. Qualifications are good, but the accolade from management and clients, "You Deliver!", is in my view more important.

Focussed on-the-job training

The most effective training in a time- and cost-constrained environment is probably on-the-job training. Here, an experienced project manager or engineer is assigned as your coach/mentor and teaches you techniques that you can apply directly to your project.

By working at the coalface, you will be exposed to both managing complex issues and delivering projects under pressure. Whilst this may be uncomfortable at times, it will help you develop as a well-rounded project engineer/manager.

PMO Summary

Project management is both an art and a science. Some things can be taught, some can be gained by experience and some come from your own personality and work style. Is experience or qualifications more valuable? Go for both, as they complement each other.

"Get the right people. Then no matter what all else you might do wrong after that, the people will save you. That's what management is all about." - Tom DeMarco

CHAPTER 18 INTRODUCTION TO THE ENGINEERING DISCIPLINES

CHAPTER 18

Introduction to the Engineering Disciplines

This third section is laid out by engineering discipline and takes you through their design processes and key interfaces and provides examples of the typical key documents they will produce. A solid understanding of the disciplines will allow you to be more effective in managing their contributions into the overall project.

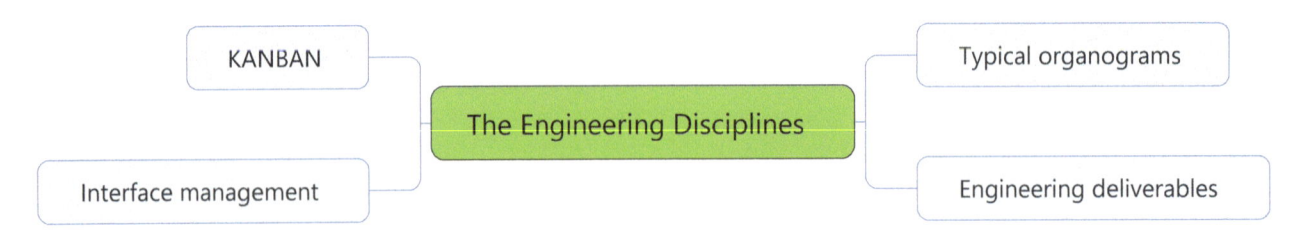

Figure 18.1 The Engineering disciplines Mind Map.

The Disciplines

The disciplines, as they are known, are the specialist engineering teams which include:

Figure 18.2 The disciplines.

As a Project Engineer, you may come from a range of backgrounds including:

PETER F CRANSTON

- A discipline engineer with a number of years of experience.
- Direct entry as a recent graduate.
- As a Project Engineer from another industry.
- As a subcontractor/supplier engineer.

Whilst you may have expertise in one of the above areas, it is not possible to have expertise in them all.

Discipline engineers normally progress from being a graduate engineer through various levels of increasing competencies and responsibilities before becoming Principal or Lead Engineers. As a Project Engineer, you will not be involved in the detailed engineering and calculations that the disciplines must perform but will have an important role in coordinating their efforts.

As the Project Engineer, you will need to have a high-level view of the overall scope of the project and be able to integrate the efforts of the disciplines. To do this, you must understand what each discipline does, their design processes, and how they interface with one another.

Tip: When joining a new project, the first two documents that I obtain are the structural General Arrangements (GAs) of the facility and the Process Flow Diagrams (PFDs). The structural GA lets me physically see what the installation looks like and the location of the main modules, such as process, utilities, drilling, and accommodations. The PFDs show the facility process at a high level.

Typical Organograms

Typical engineering organograms are shown below for a dedicated project team structure and for a matrix team structure.

Project organogram

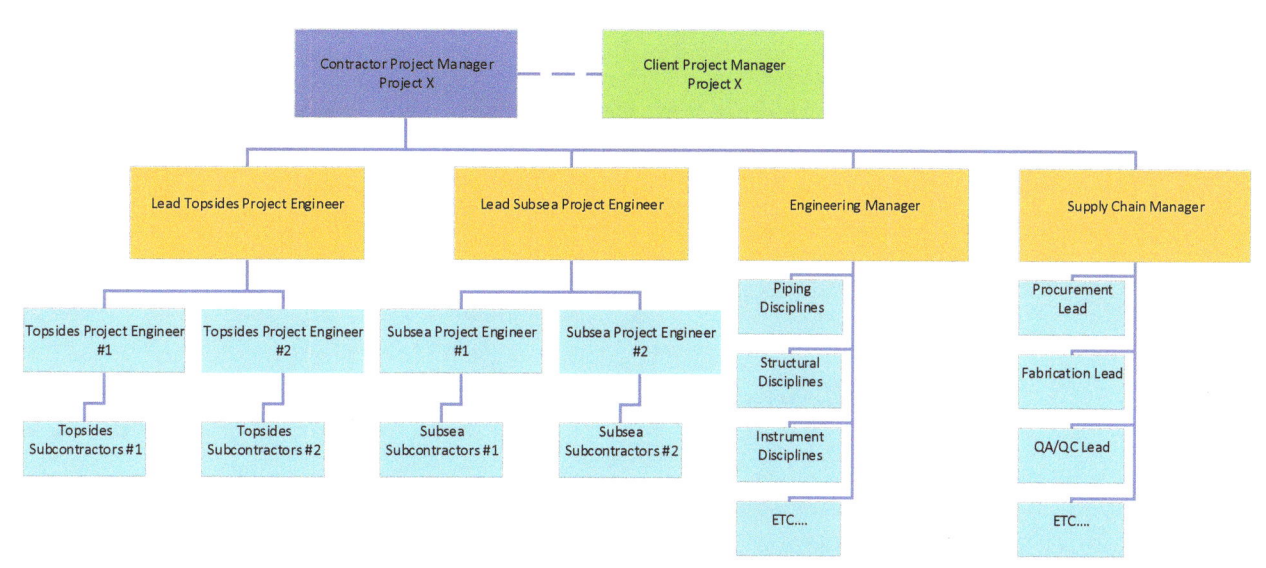

Figure 18.3 Project organogram.

Matrix organogram

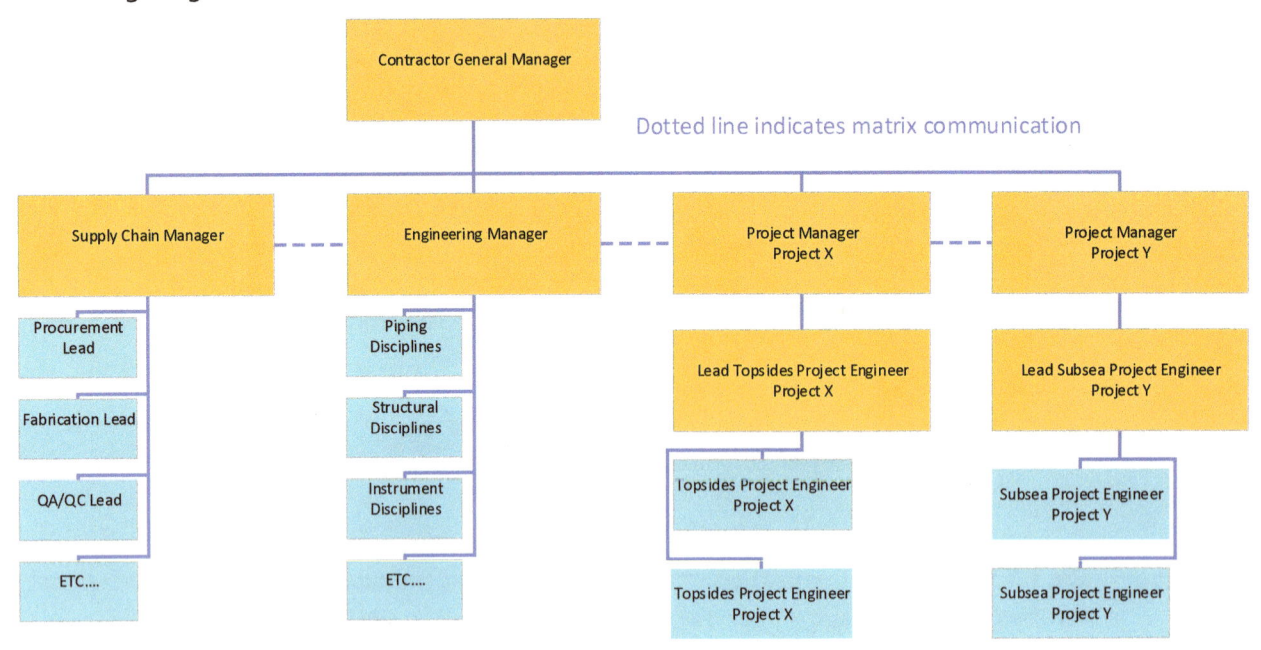

Figure 18.4 Matrix organogram.

The engineering discipline team

The engineering team will report to the Project Engineer/Project Manager concerning deliverables and scheduled dates. They will also report to the Engineering Manager on technical issues and issues of procedural compliance. The engineering manager is also normally responsible for resourcing the engineering team and up-manning/down-manning as project workload dictates.

Engineering Deliverables and Interfaces Management

The key to a successful project, especially during the design phase, is effective management of the engineering disciplines. There is a range of techniques available:

- Minutes of meetings.
- Action Logs.
- Deliverables schedules.
- Formal Interface management procedure.
- KANBAN.

Minutes of meetings

Minutes of meetings with the engineering team are useful to record interfaces and actions in detail. On the downside, though, they take significant time to prepare and update.

Action logs

Action logs using spreadsheets will not normally provide as much detail as meeting minutes but are much quicker to update and will allow all completed action history to be maintained but hidden from current view.

Deliverable schedules

Deliverable schedules are lists of what each discipline is required to produce and by when, and are normally part of the engineering plan. When sorted by due date, they allow the Project Engineer to identify any items that are forecasting late and to focus on resolving any issues associated with their completion. These issues can be technical, resource, priority, etc.

Formal Interface Management procedure

Where there are significant complex interfaces, I recommend using a formal interface procedure system that your organisation approves of or generating one yourself.

Each identified interface will have an owner and agreed interface parties. The interface will be fully defined and will include who needs what information, from whom, and by when. This normally goes through a number of iterations before all agree that the interface is complete. The keys to success with interfaces are a structured approach to interface definition, regular updates, and a formal final sign-off by all the interface parties and the Project Engineer.

Sources of interfaces

The main sources of project interfaces are:

- Sub-contractors
- Package vendors
- Area-based teams performing scope in different geographical locations.

Interface process

A typical interface process is listed below:

- Develop interface management plan and procedures.
- Identify key interfaces, define deliverables, responsibilities and dates.
- Summarise interfaces in the Project Interface Management System, extracted registers, and reports.
- Review the Interface Register at regularly scheduled interface meetings.
- Monitor, track, expedite and report on interface performance and progress.

Interface information

It is suggested that each interface sheet contains the following information:

- Interface Number.
- Revision Number.
- Revision Date.
- Title.
- Short Description.
- Data Supplier.
- Data Receiver.

- Required Return Date.
- Forecast Delivery Date.
- Close-Out Date.
- Status.
- General Comments.

A large project will require a dedicated interface manager/engineer, but on smaller projects this will typically be carried out by the Project Engineer.

Engineering Documents Process

Engineering documents and drawings are issued in a rigorously-controlled manner by the Document Control Function, (DCC). These departments have bespoke databases that track every document, to whom they were issued, when, and the revision number. Although the exact coding for documents varies with each organisation, the following generic types apply:

Issued for Review (IFR): Documents/drawings are issued for review and comment by engineers and subcontractors, with comments collated by DCC and returned to the originating engineer to review the commented document.

Issued for Design (IFD): This is where documents/drawings are sufficiently defined to allow other disciplines to carry out detailed engineering. An example to this is piping and instrument diagrams (see process chapter) that, when issued to IFD, can then be used by the piping discipline to develop detailed layouts.

Issued for Procurement (IFP): This revision allows the document to be issued to subcontractors by the procurement team and is usually very well-defined, specific in content, and unlikely to change.

Issued for Construction (IFC): These are documents/drawings that can be included in the offshore or module site construction workpacks and are used for reference whilst carrying out the offshore construction.

KANBAN

KANBAN is a visual system adapted from manufacturing for knowledge-based work such as projects. The key principles are:

- To map the work process i.e. engineering process, procurement process, etc.
- Populate a visual KANBAN board with cards indicating deliverables.
- Review regularly as a team to resolve any pinch points.

KANBAN can be highly effective in getting teams to work together and support one another. I recommend the following book to allow you to get started with KANBAN. Try it. It is very effective: *KANBAN in 30 Days* by Jannika and Tomas Bjorkholm, ISBN 978-1-78300-090-6.

I have personally implemented KANBAN on a number of projects as it reduces e-mail traffic, encourages immediate short problem-solving conversations, and improves the team spirit.

Engineering Disciplines Summary

As a PE you will only be as good as your engineering team, therefore it is incumbent on you to look after them well. What this means is being clear in your requirements, resolving issues, and providing information in a concise and timely manner.

The key to managing and supporting disciplines is communication, meaning that as PE, you take the time to filter the "noise", collate and organise information, and then output agreed actions and interfaces to the team in an easy-to-use format.

CHAPTER 19

Process

Many years ago when I was starting my Project Engineering career, an experienced Project Manager advised me of the following simple explanation of what an offshore production process does: "We drill holes, produce oil and gas, clean it up a bit, and sell it." Whilst a nice simplification, it is, in reality, much more complicated. This section will take you through an imaginary project starting with conceptual studies all the way through to commissioning and operations support.

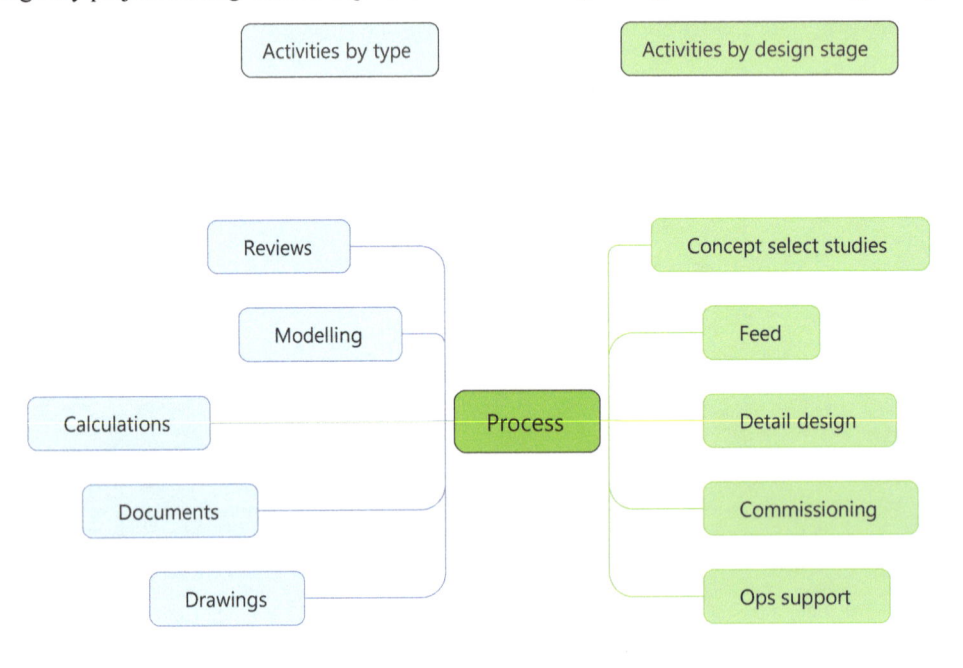

Activities by type — Reviews, Modelling, Calculations, Documents, Drawings

Activities by design stage — Concept select studies, Feed, Detail design, Commissioning, Ops support

Figure 19.1 Process Mind Map.

The Oil and Gas Production Process

All oil and gas production processes are similar. When well fluids arrive at the processing facility, they are composed of a mixture of oil, gas, and produced water. These three components must be separated into the correct specifications of oil and gas for metered export to pipelines and tankers.

The following figure shows a typical three-phase separator:

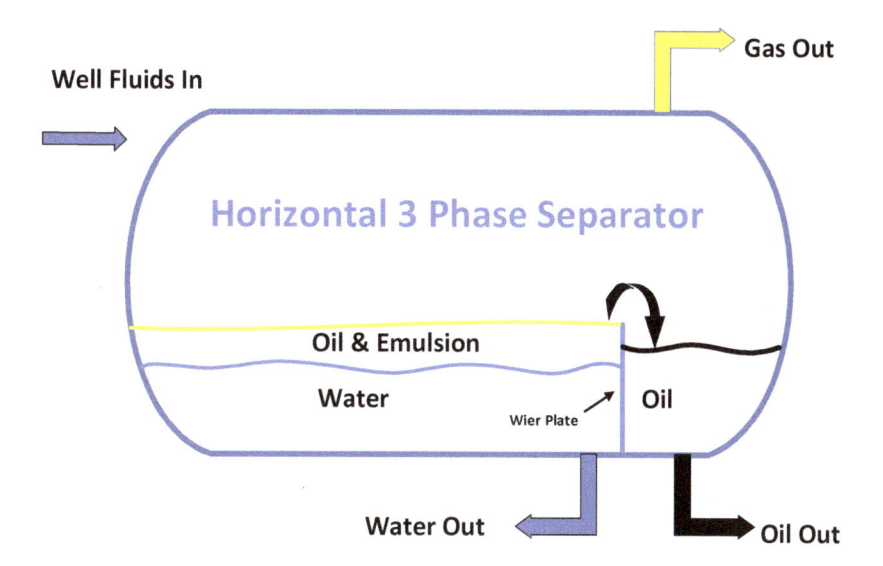

Figure 19.2 Three-phase separation.

Well fluids enter the separator where they dissociate from one another into gas, oil, and water, due to gravity and elapsed time, known as the residence time. Since water is the heaviest, it settles to the bottom of the separator and is drawn off. Oil, on the other hand, rises to the top of the liquid and spills over a weir plate, where it is then drawn off. As pressure is reduced in the separator, the gas "flashes off" and is removed from the top of the vessel. This is not a perfect separation process hence further processing is required. The following schematic shows a generic offshore installation processing configuration.

The figure shows a single production train, although it is common to have two or three trains operating in parallel to provide maximum flexibility and the ability to individually shut down parts of the system.

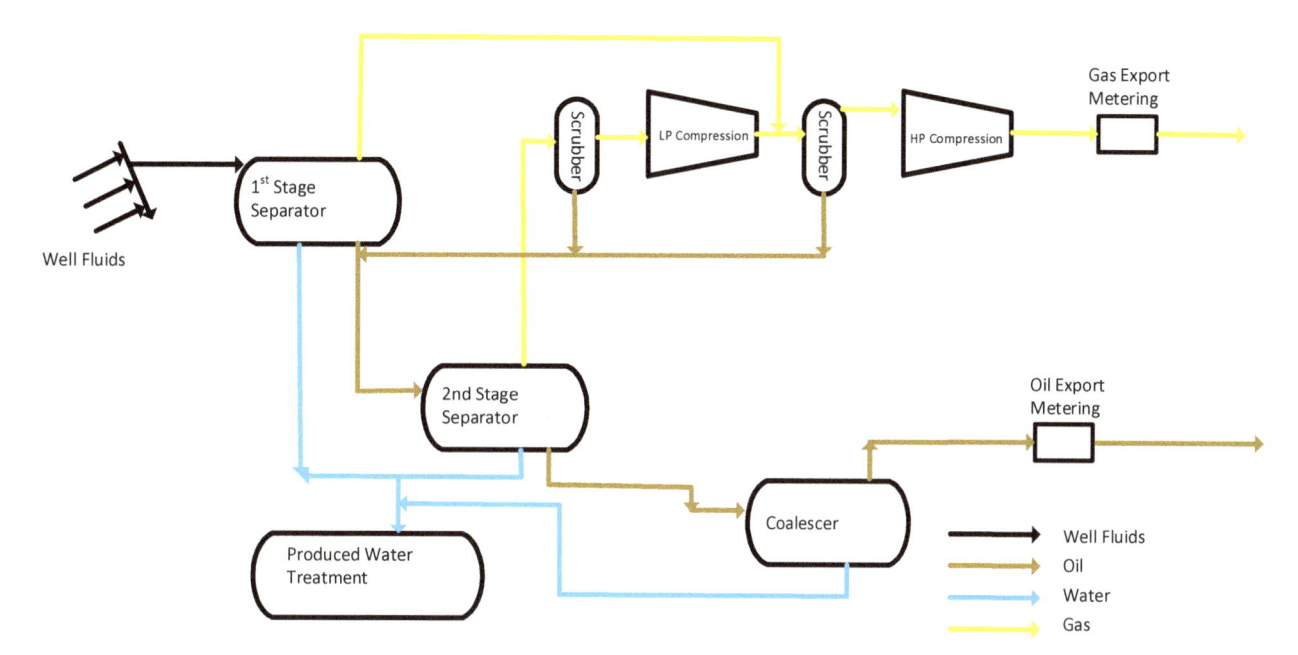

Figure 19.3 Generic production process.

The incoming well products are separated into three streams: oil, gas and water.

Oil Stream: The oil drawn off the first stage separator still has a percentage of water and entrained gas. It then goes through a second separation stage before final treatment in a coalescer to remove the remaining water. This oil is then fiscally metered before being exported to either a pipeline or tanker.

Gas Stream: The gas from the first stage separator will still have some water and oil entrained, so it must go through a scrubber to remove this liquid and return it to the second stage separator. The gas is then compressed during the high-pressure compression stage prior to fiscal metering and exported to the gas pipeline. Gas from the second stage is at a lower pressure, hence this goes through low-pressure compression before joining the high-pressure stream.

Produced Water Stream: Water from the oil and gas separation process is known as produced water. This is typically cleaned to a very high specification (30 parts per million oil in water) prior to disposal into the sea or, in some cases, re-injected back into the reservoir to provide pressure support. The most common equipment used for treating produced water and achieving very low oil in water levels uses hydrocyclone technology, which uses vortex principles to achieve separation. An internet search will provide details on the theory and practical applications.

Gated Development Process

All projects follow a series of stage gates (simplified process shown below). The actual detail will vary from organisation to organisation, but all will follow similar stages and have formal reviews and requirements to be met at each gate before being allowed to proceed to the next stage. The purpose of such a rigorous system is to ensure robustness of the design and economic justification when compared to other requests for project funding.

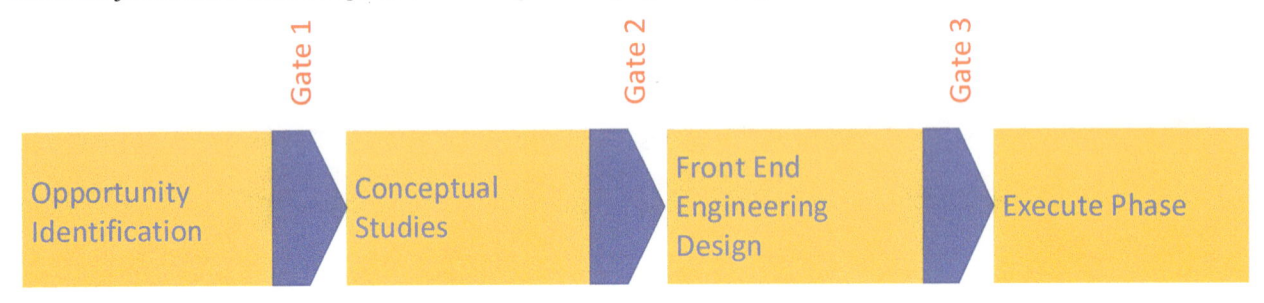

Figure 19.4 Stage gate diagram.

Requirement for Facilities

The client, usually an Oil & Gas operator, will initially identify a requirement for one of the following reasons based on a commercial opportunity, a safety requirement or new legislation:

- Complete new facilities i.e. a platform or Floating Production Storage & Offloading (FPSO).
- A new module containing process facilities.
- A significant modification to existing facilities.

Concept Select Studies

The first stage is concept select studies where a relatively small team will brainstorm and identify a range of options and sub-options based on preliminary information provided by clients.

Inputs from clients

Production profiles from client reservoir engineers

The client reservoir engineer will provide information on what the oil and gas wells are expected to produce initially and throughout the field life. They will advise production rates for oil, gas, and water.

Chieftan Gas Rates - MMscfd

Year	Existing Wells Platform	N1	N2	Total	Phase 2 Wells SW1	SW2	S3	Total	All Wells Total
2011	58	21	1	80	38	29	37	104	183
2012	55	20	2	77	24	20	27	71	149
2013	48	19	1	68	14	12	17	44	112
2014	42	17	1	60	11	10	14	35	95
2015	29	14	1	43	7	6	9	22	65
2016	29	14	1	43	7	6	9	22	65
2017	24	12	1	37	6	6	8	20	57
2018	22	11	0	33	5	5	7	17	50
2019	19	10	0	29	4	4	6	15	44
2020	17	9	0	26	4	4	5	12	38
2021	15	8	0	23	3	3	5	11	34
2022	14	7	0	21	3	3	4	10	31
2023	12	6	0	19	2	3	4	9	27
2024	11	6	0	17	2	2	3	7	24
2025	10	5	0	16	2	2	3	7	22

Table 19.5 Typical production profile

Compositions

The reservoir engineer will also provide compositions of the well products gained from exploration drilling or test production on existing facilities. This identifies the various hydrocarbon compounds and their expected percent content. It will also identify any other potentially unwanted produces such as hydrogen sulphide (H_2S) which will have to be dealt with in the production process.

Chieftan Reservoir Fluid Composition - Water-free Basis (mol%)

Component	Well N1 Liquid	Well N1 Gas	Well N1 Combined	Well N2 Liquid	Well N2 Gas	Well N2 Combined	Average	Well 22/9-4
CGR	40 bbls/MMscf			34 bbls/MMscf				
Hydrogen	0	0	0	0	0	0	0	0
H_2S (Note 1)	0	0	0	0	0	0	0	0
Carbon dioxide	0.40	2.58	2.55	0.18	1.10	1.07	1.81	2.41
Nitrogen	0.08	1.89	1.87	0.04	1.42	1.38	1.63	1.58
Methane	6.91	81.35	80.47	7.52	86.72	84.42	82.45	79.60
Ethane	2.22	7.07	7.01	2.22	6.05	5.94	6.48	6.87
Propane	2.87	3.63	3.62	2.67	2.60	2.60	3.11	3.57
i-Butane	0.90	0.55	0.55	0.88	0.39	0.40	0.48	0.58
n-Butane	2.74	1.19	1.21	2.61	0.81	0.86	1.04	1.31
neo-Pentane	0.01	0.01	0.01	0.01	0	0	0.01	0.50
i-Pentane	1.81	0.34	0.36	1.80	0.22	0.27	0.32	0.59
n-Pentane	2.87	0.42	0.45	2.89	0.27	0.35	0.40	0.50
Hexanes	6.34	0.38	0.45	6.54	0.2	0.38	0.42	0.59
Me-Cyclo-pentane	1.74	0.07	0.09	1.71	0.03	0.08	0.09	0
Benzene	0.84	0.03	0.04	1.43	0.03	0.07	0.06	0

Table 19.6 Typical compositions

Production specifications

The client will provide the specifications that the processed oil and gas must meet before it is exported to pipelines or tankers. This will include maximum and minimum flowrates, temperatures, water content, H_2S content, etc.

Conceptual options

Armed with the above information, an experienced studies team will generate a range of options from which they will produce the following key information:

- System production capacity and expandability.
- Production availability (percent of time it is expected to be available to produce).
- Screening costs.

Generate PFDs

For each identified option, the team will produce a Process Flow Diagram, (PFD) that allows for completion of preliminary mass balances and identification of major equipment. Most usefully for the Project Engineer, it provides a simplified overview of the process.

Tip: It is worth getting the PFDs for the asset or project you are working on and following the process from wellhead/riser to export and, if offshore, walking the route in-person.

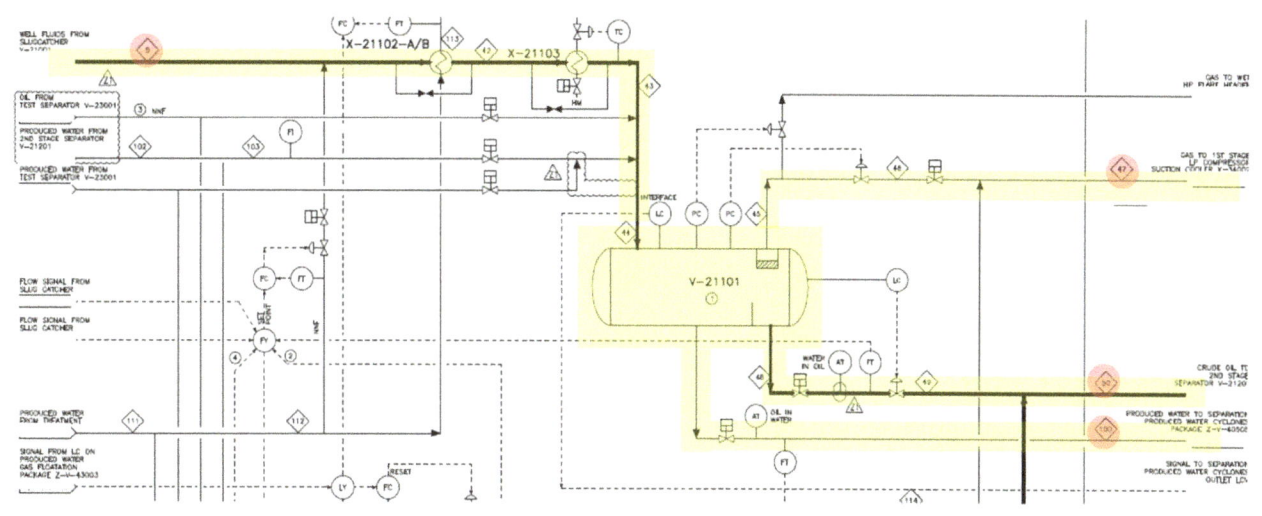

Figure 19.7 Example PFD excerpt.

Assess and compare development options

As the team develops these options they will regularly review them both within the team itself and with the client. It is normal during this process for some options to be discounted and other options/sub-options to be generated.

Example Options:
Fixed platform or FPSO, two production trains, or three production trains, etc.

Fixed Jacket Platform	Yes	Yes	No
FPSO	No	No	Yes
No of separation trains	2	2	3
No of stage operating pressures	2	3	3
Produced water treatment	No	Yes	Yes
NGL recovery	No	Yes	Yes
Export oil cooling	No	No	Yes
Etc.			

Table 19.8 Example table of options.

Identify major equipment items

The study team will also identify major equipment items to allow preliminary costs to be generated. This will include:

- Pressure Vessels.
- Pumps.
- Compressors.
- Heat exchangers etc.

Preliminary heat and mass balance

Using the PFD and production profiles, the process engineers will complete preliminary heat and mass transfer balances for all streams into and out of the proposed production process. These calculations are used to balance all the mass inputs of the system with the outputs, wherein the net energy change in the same system should be zero (i.e. heat and mass balance).

Operability review

As the study develops, the operability of each option will be evaluated, most likely with client operations input. They will look at availability, redundancy, and flexibility of operation.

Conceptual study report

The output of the conceptual study is the conceptual study report that is submitted for review, commented on, and approved for wide distribution. The conceptual study report will normally have a recommendation on the preferred option to take through into FEED.

Stage gate review

Each client has their own review and approval procedures and will submit the provided conceptual study output for approval to move to FEED.

FEED

Front-End Engineering Design (FEED) will normally be approved to move the selected option forward, sometimes with some sub-options requiring refinement. At this point, the team size will increase alongside the cost expenditure. The purpose of the FEED is to develop the design to a sufficient level of detail for enabling scope, costs, schedule, and risks for the project to be approved into the execution phase i.e. to complete the detail design, procure equipment/materials, and construct and commission the project.

Process Basis of Design (BOD)

This document is an input from the client to the engineering contractor process team, which represents the client's instructions to the design engineering organisation about what the owner wants designed and how the owner wants the process design to be done. The design basis also provides the who, how, and why the facilities are needed. It typically includes the design philosophy, the design standards, and the format of deliverables.

Through field life heat and mass balance

The team will revisit the preliminary heat and mass balance and go into more detail for production throughout the field life.

PFD & UFDs

More detail will be added to the PFD. In addition, utility flow diagrams (UFDs) will be generated for the supporting utilities, including cooling water, compressed air, heat transfer fluids, etc.

HYSYS modelling and calculations

Process system modelling will be carried out using HYSYS™, a chemical process simulator used to mathematically model the production process shown on the PFD. HYSYS™ is able to perform many of the core calculations of process engineering, including those concerned with mass balance, energy balance, and pressure drop. Sensitivities will also be carried out to allow the process to be optimised.

Outstanding concept selection issues

During concept there may not be enough time or information to fully define everything, so during FEED any outstanding concept select issues will be resolved. For example, whether to install two or three stage gas compression.

Pipe sizing

As definitions improve and pressures and flowrates are defined, pipe sizing can be established.

Revisit equipment sizing

Equipment sizing can be revisited based on the latest and most accurate information.

Safety relief valve sizing

As the process design develops in FEED, the process safety valves can be sized for setting ranges and flow capacity.

Optimisation

As FEED is completed, the process team will optimise the design, trading-off various features and limitations for others.

Output of the FEED process

The process team will complete the following key deliverables which will be used by the other disciplines to complete their own inputs into the overall FEED report.

- Approved for Design process and instrumentation diagrams.
- FEED HAZOP.
- Safety Integrity Level (SIL).
- Key vendor interfaces.
- Equipment data sheets.
- Process line list.
- Operating philosophies.

Approved for Design process and instrumentation diagrams

Approved for Design (AFD) Process and Instrumentation Diagrams (P&IDs) detail the entire process with every vessel, pump, heat exchanger, instrument, pipe specification, and size detailed.

FEED HAZOP

The hazard and operability study (HAZOP) is a structured and systematic examination of the proposed process as detailed in the P&IDs. The purpose is to identify and evaluate problems that may represent risks to personnel or equipment.

Safety Integrity Level (SIL)

Safety integrity level (SIL) is defined as a relative level of risk-reduction provided by a safety function or to specify a target level of risk reduction. In simple terms, SIL is a measurement of the performance required for a safety instrumented function (SIF). Normally, four levels of SIL are defined, with SIL 4 being the most dependable and SIL 1 being the least. From a project perspective, SIL 4 offers the greatest risk reduction although it usually comes with greatest complexity and cost to both engineer and purchase.

Key vendor interfaces

During the FEED there will be interfaces established with key providers of equipment to establish the availability and limitations of that equipment. It is common for long lead-time items to be ordered in the FEED to reduce the overall project duration, although this does introduce its own risks.

Equipment data sheets

A process data sheet will be provided for each item of equipment to allow the procuring package engineers to fully specify requirements.

Process line list

A key deliverable from the FEED is the process line list that details the specification of each process line, i.e. materials specification, required pressure rating, outside nominal diameter, and design maximum and minimum operating pressure and temperature.

Rev	Size/NB	Service	Line Number	Spec.	Ins.	P&ID Number	Material	Fluid	Fluid Phase	From	To
	8"	OC	70638	1C1B1	-	SK-MAPG-L-0030/010-D-AC00-0004 (0426-110-K010-PID-002-001)	LTCS	Crude Oil	L	BLP Oil Export 8"-OC-80929-4C1B1	MOL Pump Common Suction Header 12"-OC-70701-1C1B1
	12"	OC	70701	1C1B1	-	SK-MAPG-L-0030/010-D-AC00-0004 (0426-110-K010-PID-002-001)	LTCS	Crude Oil / Seawater	L	8"-OC-70638-1C1B1 12"-OC-70765-1C1B1	8"-OC-70702-1C1B1 8"-OC-70706-1C1B1 8"-OC-70710-1C1B1

Figure 19.9 Example line list excerpt

Every process line is included on the line list and includes the following information:

- Nominal bore in inches.
- Fluid service.
- Line number.
- Piping specification.
- Insulation requirements (if any).
- P&ID number.
- Material.
- Operating pressure and temperature.
- Design pressure and temperature.
- Test pressure.
- Specific comments to the particular line.

Operating philosophies

Operating philosophies are usually developed with client operations personnel input advising on how the equipment is to be operated.

FEED summary

The FEED is probably the most important stage of the project. Its importance cannot be underestimated. The oil industry is littered with project disasters in which not enough time was allotted for or enough effort put into a quality feed. The cost and schedule impacts of changing a design after detail design can be huge, negatively impacting reputations and client relationships.

Detail Design

During detail design, the team size will increase as there are many detailed calculations and specifications to be developed based on the FEED report.

AFC P&IDs

Detail design refines and updates the P&IDs until they are Approved for construction (AFC), i.e. the other disciplines can start procuring equipment and materials.

Complete line list

The line list developed in FEED is updated to 'Approved for Construction' to allow piping design and drawings to be completed and piping procured and fabricated.

Detail design HAZOP

The FEED HAZOP is revisited using the AFC P&IDs. Hopefully minor issues will be identified at this stage which will be iterated through the design process.

Locked valve register

This is a register which details all valves that will be either locked open or closed during operation.

HP/LP interface register

This is a register which details where there are changes in pressure ratings of systems, from low pressure to a high pressure and vice versa. These are important to ensure that conditions do not arise where a system is over pressurised with potential loss of containment.

Update calculations

Calculations done in FEED are refined with more accurate information.

Reviews

A number of reviews will be completed including SIL, Environmental Impact Identification ENVID, etc.

Disciplines interface

The key users of the process design output information are piping, instrumentation and control who will develop their disciplines design to AFC stage.

Detail design summary

In summary, detail design is as FEED with improved definitions and removal of design options

Commissioning

As detail design is completed, a much smaller process team will normally become part of the commissioning and start-up team that will support the commissioning team in development of the commissioning procedures. Oil and gas production processes rarely work properly at start-up and one or two of the more experienced process engineers will travel to the site to support commissioning and provide on-site expertise. Full details on commissioning are provided in Chapter 31, "Commissioning".

P&ID As-Built

As each system is commissioned, there may be minor changes to the configuration of equipment and valves which are marked up in red pen on the site as As-Built drawings. When these "red pen" mark-ups are returned onshore, the drawings are "back-drafted" and issued formally as As-Built. This allows all subsequent P&ID users on and offshore to have accurate information.

Operations Support

A significant number of process engineers are involved in operations support, which involves day-to-day optimisation of the production process. These process engineers will be involved in improvement projects that do not require an engineering contractor's resources, and they can be advised of a process issue in the morning and be expected to advise corrective or mitigation actions by the next day.

Operations process engineers are primarily reactive. This can be real "seat of the pants stuff" and is best suited for those who thrive in a high-stress, rapid response environment.

Documents maintained by operations process engineers

- P&IDs.
- Line lists.
- Operating procedures.
- Locked valve register.
- HP/LP Interface Register.

Drawings and Documents Produced by the Process Discipline

The following list details the documents and drawings that are produced by Process. While you do not need to be able to do the calculations, you should be able to follow the information provided in these documents and drawings at a high level.

Drawings

- PFDs.
- UFDs.
- P&IDs.

Documents

- Line lists.
- Data sheets.
- Basis of Design BOD.
- Technical notes for decision support.
- Reports.
- Operating philosophies.
- Operating procedures.
- Calculations.

Calculations

Standard black box spreadsheets

These are either engineering-, contractor-, or supplier-developed spreadsheets that allow calculations for frequently used process items such as:

- Pipe sizing.
- Control valve sizing.
- Relief valve sizing.

Bespoke spreadsheets

Bespoke spreadsheets may be developed for specific systems that allow sensitivities to be easily and quickly carried out.

Subcontract flow assurance

Flow assurance modelling is a specialist activity that is often subcontracted. It refers to ensuring a successful and economical flow of hydrocarbon stream from reservoirs to the point of sale. Flow assurance is extremely diverse, encompassing many specialised subjects, including:

- Thermal investigation and risk of hydrate formation (ice-like solids that block pipelines).
- Modelling of pipelines and risers flow.
- Slug suppression.
- Determining fluid arrival conditions at the process facilities.

Process modelling

The most common types of modelling carried out are:

- HYSYS™: Process system modelling.
- Flarenet™: Flare system modelling.
- Subcontract Computational Fluid Dynamics, CFD.

Routes to becoming a process engineer

The entry route to becoming an Oil and Gas process engineer is a degree in chemical engineering. Most process engineers start off as graduate engineers and build experience over a number of years, supported by additional specialist courses.

Process Summary

Process is normally the lead discipline on a project involving process scope. The other disciplines then develop their scope using process output information such as P&IDs and Line Lists. The process should always give sufficient time to develop the design before the other disciplines commence as fast-tracking other disciplines in parallel will introduce risk, cause confusion, and the need for engineering rework.

CHAPTER 20

Piping

The piping discipline is often the largest sub-team on a project and is generally responsible for three areas: layouts spatial management, piping design, and procurement of correct piping materials.

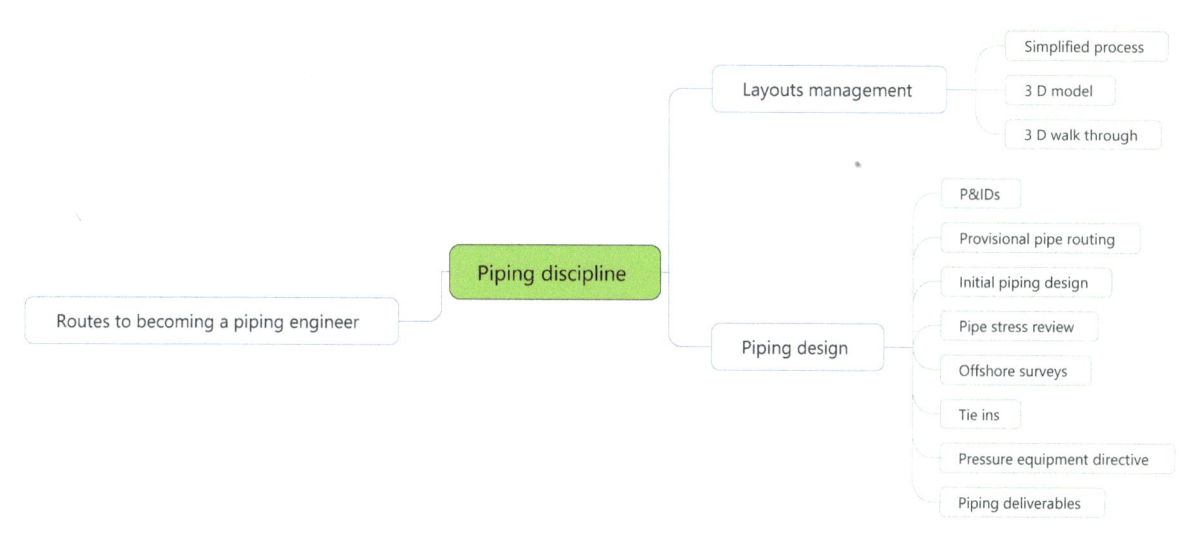

Figure 20.1 Piping Mind Map.

Layouts Management

On larger projects there is a requirement for space management of all the equipment and services of a particular module or facility that is managed by the piping discipline using 3-D CAD software, such as AVEVA PDMS™ or AutoCAD Plant™.

Simplified process

The simplified design process is as follows:

- Identify what overall space is available for the equipment and services.
- Locate all the major vessels and equipment packages within the overall space.
- Route the major pipework and locate major valves.
- Locate escape routes and walkways.
- Route secondary pipework.
- Locate instruments.

- Locate cable trays (by electrical and instruments disciplines).

3-D Model Reviews

At regular stages there will be multidiscipline layout reviews that will typically use a live 3-D model on a large screen to allow the team to "walk-through" the module, resolve issues, and determine optimum solutions. Any issues highlighted will be noted and added to an action tracker to ensure they are closed out in a timely manner.

Reproduced by permission of CNOOC Ltd

Figure 20.2 Typical 3-D model picture.

During these iterative reviews, the following will be checked and refined:

- Overall weight/centre of gravity and impact on primary structure.
- Access and egress routes, including escape routes.
- Operations access.
- Maintenance access including location of lifting beams.

Reviews are typically carried out at the following levels of definition:

- 30% scope overview inputs and outputs.
- 60% scope overview inputs and outputs.
- 90% scope overview inputs and outputs.

3-D walkthrough

The output from layouts is a controlled model of the module or facility which will be passed over to the client upon completion of the project. This 3-D model will be maintained and used when evaluating any future engineering modifications.

Tip: A word of caution: Many 3-D models have not been 'As-Built' or properly maintained since the initial project, resulting in discrepancies with the real-life structure. Designs should be backed up with physical offshore verification surveys to ensure accuracy and avoid expensive clashes or rework when constructing or using information from a point cloud survey. Point cloud is a specialised laser scanning process that builds a true picture of the surveyed site and is recommended to ensure the model is up-to-date.

Some clients and main contractors have dedicated 3-D cinemas with large curved screens where a fully immersive walkthrough can take place when wearing 3-D glasses. This is particularly useful when engaging the plant operators who will ultimately use the facility to enable the incorporation of any specific changes prior to any fabrication. These 3-D models can also be used to generate recorded videos showing installation and construction sequences of the offshore facility that are great for explaining the project scope to the wider teams and stakeholders.

Piping Design

For the purposes of this book please assume that we are installing a 1000-tonne process module on an existing platform to illustrate both the greenfield and brownfield aspects of the process.

The piping team will require the following information to get started:

- Approved for Design (AFD) and P&IDs from the process team.
- Line lists from the process team.
- Equipment lists from the mechanical team.
- Any previous study or FEED reports including:
    - Existing 3-D models.
    - Offshore survey data.
    - Instrument in line valves and actuators ESDV's, PSV's etc.
    - Material selection report from metallurgy.

P&IDS

The P&IDS show the schematic connection of piping between items of equipment. They do not show lengths or routes.

Line lists

The line lists contain the following information:

CHAPTER 20 PIPING

- Pressure ratings.
- Materials specifications.
- Design pressures.
- Maximum and minimum design temperature.

Provisional piping routing

Provisional piping routing is carried out in the FEED stage allowing for long lead piping material to be ordered, sometimes with lead times of many months.

Reproduced by permission of CNOOC Ltd

Figure 20.3 Existing layout modelled in 3D CAD

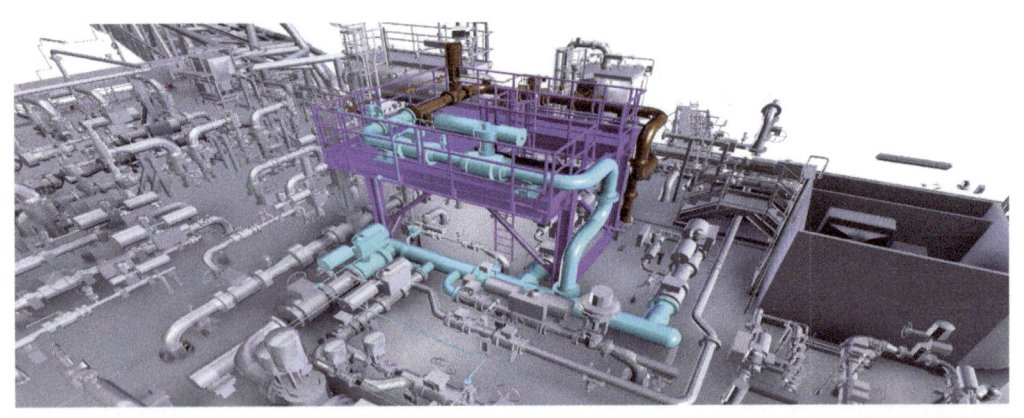

Reproduced by permission of CNOOC Ltd

Figure 20.4 Proposed modified layout modelled in 3D CAD

PETER F CRANSTON

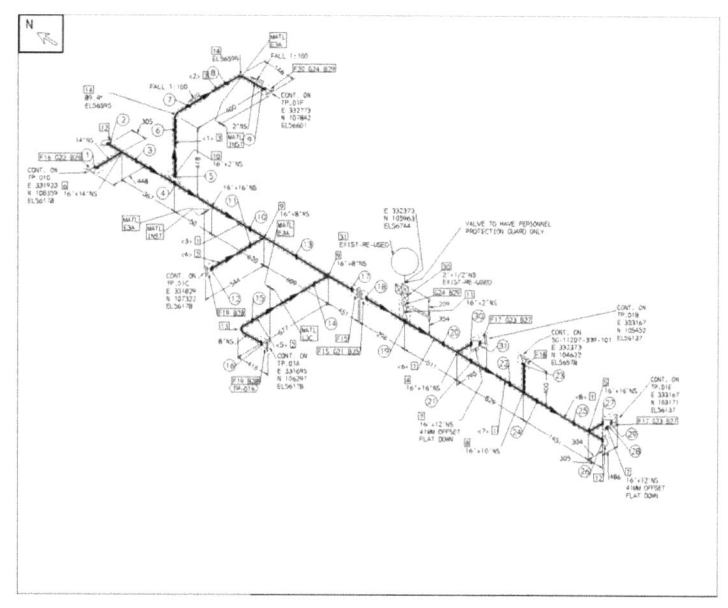

Reproduced by permission of CNOOC Ltd

Figure 20.5 Fabrication isometric drawing

Armed with this information, the piping designers can then route pipework within the 3-D model. Modern plant design 3-D software is intelligent and contains information relating to each piping type specification, allowing for partially automated generation of bills of materials and other information.

Initial piping design

When sufficiently developed, an offshore dimensional survey is completed to determine exact locations of the piping connection tie-in points to any existing equipment and to confirm that the piping and equipment will fit without clashes. The dimensional survey is carried out by specialists using laser survey equipment that provide high levels of accuracy to within +/- 1 millimetres of tolerance.

Pipe stress review

Piping systems are required to cope with a range of stresses including self-weight, environmental, thermal and, in the worst-case situation, explosion overpressure loads. This analysis is carried out by specialist pipestress engineers, typically using Caesar II™ software, who will then advise of the loads, piping geometry changes, and types of pipe supports required.

Design Survey

For brownfield projects, offshore surveys are typically carried out by piping designers. There is nothing more effective for a designer than walking round the worksite in-person, making sketches, taking notes and photos, and then working out the piping routes on-site. Armed with this information, 3-D modelling onshore can proceed with a very clear understanding of routing options and the offshore limitations. It should be noted that 3-D models always look less congested than the real offsite worksite where over time things will have been added or modified.

Specialist dimensional surveyors carry out two types of survey on behalf of the piping team:

- Laser point-to-point surveys that are normally completed using the R1, (for comment), isometrics and to verify dimensions.
- Point cloud surveys scanning an entire area providing details to validate the 3-D model.

Once offshore dimensional surveys and pipe stress have been completed, the piping supports can be designed and detailed. Looking to the future there are a number of technology advancements using tablets, handheld scanners, and measuring devices currently coming to the market which will further enhance offshore design surveys.

Tie-Ins

On many brownfield projects there will be a requirement to provide Tie-Ins to the existing system prior to the main scope being installed. This is usually done in an opportune or planned process shutdown where double-block and bleed isolation valves are installed.

Pressure equipment directive (PED)

Legislation regarding pressurised system varies, but in the European Union (EU) there is a requirement for compliance with the Pressure Equipment Directive (PED). PED applies to the design, manufacture, and conformity assessment of pressure equipment and assemblies with a maximum allowable pressure greater than 0.5 bar gauge to ensure common standards of safety in all pressure equipment sold within the European Economic Area.

Piping deliverables

The final output from the piping team at the end of detail design will include:

- A clash-free design.
- Piping C1 fabrication isometrics drawings.
- Pipe supports C1 fabrication drawings.
- Bills of materials for bolts, gaskets, etc.
- Material requisitions for manual valves, pipe fittings and flanges, special piping items, bolts, and gaskets.
- Review and approval of materials certification for all items purchased by piping.
- Piping stress analysis report.
- Pressure Equipment Directive (PED) file.

Routes to Becoming a Piping Designer/Engineer

Many piping designers start their careers as junior designers/apprentices and study part-time. Those with the aptitude and appetite can progress to piping engineers, often obtaining part-time degrees to support their development. Direct entrants as piping engineers are likely to have gained a degree in a mechanical engineering or similar.

Depending on aspirations, the routes within the piping discipline can lead to specialisation in layouts, materials, or stress engineering.

PETER F CRANSTON

Piping Summary

As mentioned at the start of this chapter, the piping discipline is often the largest sub-team and will therefore be one of the greatest design costs. Hence, ensuring optimal information flow through this team and removing bottlenecks will have a significant impact on the overall design costs.

CHAPTER 21

Mechanical

Mechanical engineering covers a very wide area of specialist areas depending on the organisation. In main engineering contractors for offshore facilities, the mechanical discipline is responsible for management of mechanical equipment packages from initial specification, manufacture, assembly, testing, and installation through to final commissioning and handover offshore.

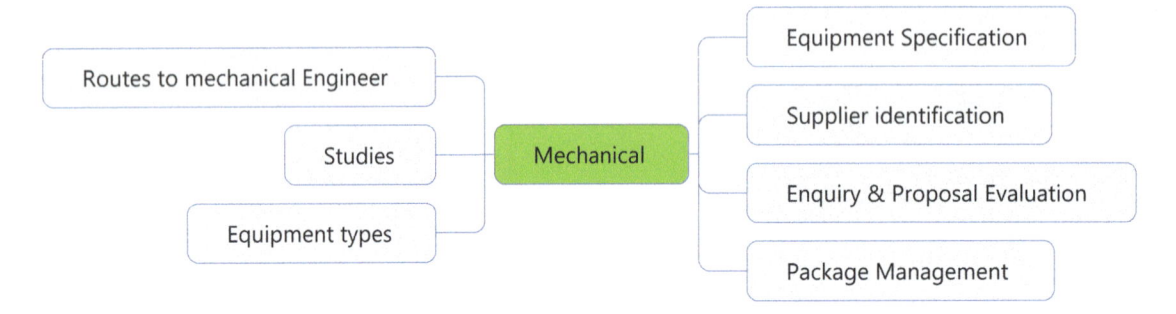

Figure 21.1 Mechanical Mind Map.

This chapter will take you through the package management process and then look at some specifics of the mechanical engineering discipline.

PETER F CRANSTON

THE PROJECT ENGINEER'S TOOLKIT•173

Reproduced by permission of Repsol Sinopec Resources UK Ltd

Figure 21.2 Typical mechanical package: Seawater filter.

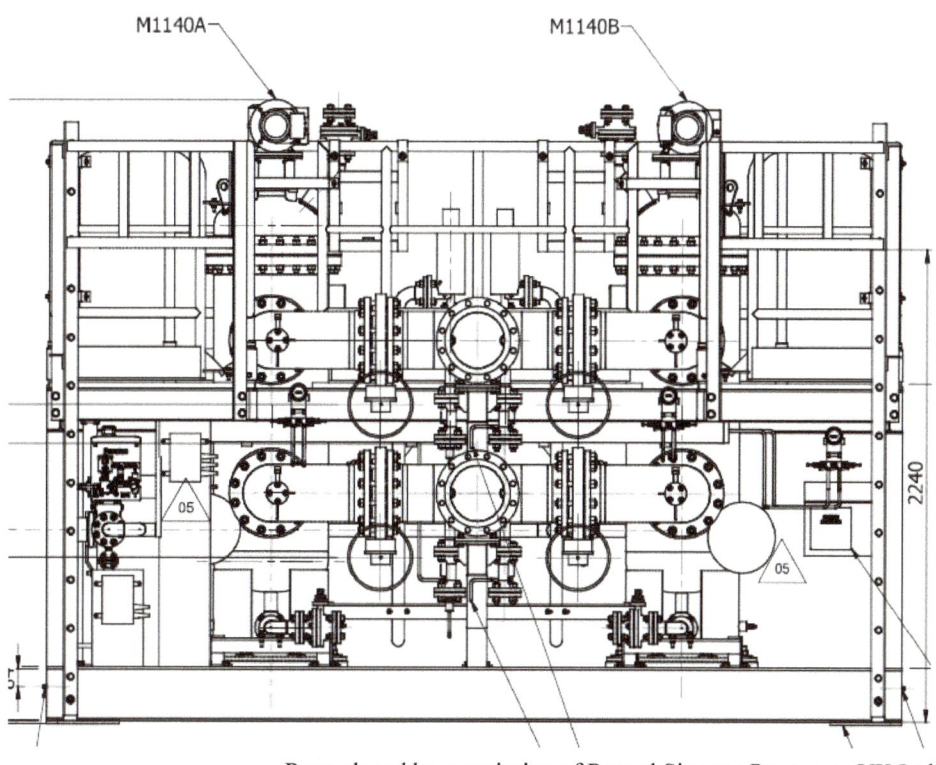

Reproduced by permission of Repsol Sinopec Resources UK Ltd

Figure 21.3 Seawater filter general arrangement

The Package Management Process

Packages can range from simple and small pump skids to complex gas turbine driving compression packages. However, the management principles are similar.

Package Management Process → Equipment Specification → Supplier Identification → Enquiry Proposal Evaluation and Award → Package Management → Installation and Commissioning

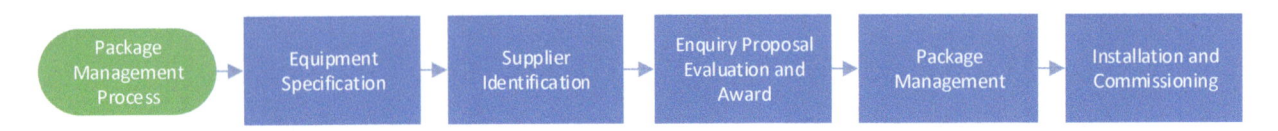

Figure 21.4 Package management process.

Equipment specification

Identification of requirements

The process starts with the mechanical engineer initially gathering requirements for the package using:

- Client specifications & standards.
- Legislative requirements.
- International and national standards.
- Information from the other disciplines.
- Information from site surveys (for brownfield projects).

Criticality assessments

A formal evaluation of the criticality of the equipment must be carried out to determine the level of quality assurance that will be required.

Data Sheets and Equipment Specifications

The mechanical engineer will then prepare equipment data sheets, specifications, and the supplier document requirements list. The mechanical data sheet contains the following items:

- Process data.
- Design and construction codes.
- Materials of construction.
- Site conditions and locations.

Equipment requisition

This package of information is then attached to the requisition which will be issued in procurement to allow the formal enquiry process to be carried out. The requisition contains requirements for the following items:

- Engineering documents.

- Interface documents.
- Manufacturing documents (welding and testing, material certificates).
- Site preservation and packaging.
- Commissioning procedures.
- Manufacturing Record Book and operations and maintenance procedures.

Supplier Identification

Potential supplier identification is normally an iterative process. Initial identification of potential suppliers can come from the organisation's approved vendor list, the client's approved vendor list, r from industry databases such as First Point Assessment List (FPAL). For more complex and expensive packages, suppliers will be invited to qualify for tendering via a questionnaire that covers technical, commercial, and safety performance. Mechanical engineers will input the technical section of this pre-qualification questionnaire.

Enquiry and Proposal Evaluation

On receipt of the pre-qualification questionnaire, supply chain will contact potential suppliers in order to pre-qualify them. This short list of pre-qualified suppliers will then be used by the procurement department when issuing the formal enquiry.

Supply chain single point contact

It is important to note that during the enquiry and evaluation phase, all contact with suppliers goes through supply chain in order to protect all parties and prevent claims of favouritism or improper actions. Supply chain will collate the technical and commercial responses from each supplier and will issue the technical sections only to mechanical engineering. Commercial information will be retained by supply chain to avoid any commercial influences on the technical decisions.

Proposal distribution and review

The mechanical package engineer will then distribute the various technical sections to other appropriate disciplines, such as electrical and instruments for review and comment. The purpose of the review is to confirm compliance with the technical requirements of the requisition and to identify any instances of non-compliance. The package engineer will then collate all the comments and issue clarifications to supply chain to pass on to the suppliers.

Bid conditioning

Bid conditioning, if required, ensures bids are fully responsive to all the requirements that are specified in the enquiry documents and can therefore be evaluated on a like-for-like basis. Bid conditioning can include minor changes to specification when advantageous and provide better value for money to the procuring organisation.

Clarifications and technical bid evaluation

After a number of rounds of clarifications and deviation requests, all answers should have been provided for and the package engineer can complete the technical bid evaluation (TBE) and issue a formal report that will advise which suppliers have a technically compliant proposal.

Award recommendation

Supply chain will then review the commercial proposals from technically compliant suppliers and recommend award based on project cost, schedule, and contractual requirements.

Documentation for purchase

The package engineer will then revise the documentation for purchase, reflecting the agreed purchase requirements and reissue to the supply chain department to allow contract award. Depending on the complexity of the package, it can take many weeks or months to go from initial enquiry issue to order award.

Tip: Many mechanical engineering packages are long leads in a project and may require more engineering effort earlier on. Ensure that procurement enquiry and lead times are adequate and still meet overall project schedule requirements.

Package Management

Package engineers are responsible for technical delivery of their allocated packages, which is achieved through a very structured and rigorously controlled process. The following steps are typically followed for all packages:

Kick-off meeting

This is normally carried out at the supplier's premises to ensure maximum supplier personnel attendance.

Pre-inspection meeting

This meeting allows the inspection team and supplier to be fully briefed on the required inspection requirements.

Supplier documents review and approval

The package engineer will receive and distribute supplier documents to other relevant disciplines seeking comments required to obtain approval. Mechanical engineering will review, collate, and code these comments before returning them to the supplier. Coding will be selected from one of the following:

Code	Description of code
1	Not Approved
2	Approved Subject to Comments
3	Approved
4	For Information Only

Table 21.5 Typical supplier documentation coding.

This formal control of the supplier's documentation at each stage, combined with Quality Assurance (QA) and Quality Control (QC) will ensure that the specified equipment will be delivered.

Concession requests

During equipment manufacturing, it may not be physically possible to achieve compliance with a purchase order requirement specification without impacting the overall project and a concession request that will identify non-compliance will be raised. This will then be formally submitted to the client for acceptance. The package engineer will review, potentially with other specialists and technical authorities, and then decide whether to accept the concession. This may happen multiple times in very complex packages. An example would be a minor change in a valve specification to allow a shorter lead time valve to be procured from a sub-supplier.

Progress meetings, inspection visits. and witness testing

The package engineer and his/her supporting QA and QC team will regularly visit suppliers to determine progress and instruct any remedial actions. Package engineers typically do a lot of international travel to the various suppliers.

Factory Acceptance Testing (FAT)

FAT testing will be carried out in accordance with agreed procedures and attended by the package and QA engineers. Operators will normally attend the FAT and be part of the commissioning team as they take ownership of the new equipment.

Equipment release

On completion of the factory acceptance testing, (FAT), and associated documentation, the package engineer will formally release the package to be delivered to the intended worksite or into storage.

Training and final manuals

To ensure that the facility client operators can safely and efficiently operate the new equipment, they will be provided with the manufacturers' equipment-specific training and operating manuals.

Transport and storage

Transportation, especially for large items, must be planned in advance and may require quayside load-out, police road escorts, and specially-designed lifting beams. Many packages will have specific transportation and storage requirements such as disconnection of couplings, and filling with inert gases and preserving fluids. Proper preparation will minimise risk of damage due to extended storage.

Certification data book

The full certification and data book normally follows delivery and will contain all material and testing certifications, such as Certificate of Conformity and other documentation as specified in the requisition.

Installation and commissioning

The package engineer will provide support during installation and commissioning, often located on-site. Depending on the complexity of the equipment, the supplier may have a supporting engineer on-site during key phases.

Mechanical Engineering Specifics

Equipment types

There are five main equipment types that mechanical package engineers are involved in, and it is common for mechanical engineers to specialise in one of these fields:

- Equipment packages.
- Rotating equipment.
- Pressure vessels.
- Cranes & lifting equipment.
- Proprietary/specialist equipment.

The specialist engineers in each of these areas will be knowledgeable on the application of industry and client codes, and standards relating to the particular type of equipment, and will complete out the required design calculations and associated specification documents.

Equipment interfaces

Equipment packages are usually specified to fit within predetermined space and weight envelopes allowing the platform or module layout to be designed concurrently with the package by the subcontractor. The platform P&IDs will normally show equipment packages as "black boxes," finishing at the various piping and instrumentation tie-ins. The equipment subcontractor will then produce the detailed P&ID for their equipment with external connections mirroring those on the platform P&ID.

The subcontractor general arrangement (GA) drawings will show overall dimensions, weights, and the exact locations of connecting pipework, foundations, and lifting points.

Packages

The term "package" refers to any procured equipment that requires oversight through design and delivery by an engineering discipline. A mechanical package can be defined as a composite assembly of components i.e. pressure vessels, pumps, motors, pipework, valves, and controls that performs a required function, such as a glycol dehydration unit. The package dimensional limits and tie-in points will be specified to allow installation into the allocated space and connection of piping, electrical, and instrumentation systems.

The package engineer will be responsible for the technical delivery and coordination of all internal and external specialists required to design, manufacture, and test the unit.

Rotating equipment

This covers:

- Gas turbines.
- Compressors.
- Pumps.

Rotating equipment engineers will prepare detailed requirements regarding specifications, testing, and materials of manufacture, as well as input into reliability and availability evaluations of the overall process.

Pressure vessels

A pressure vessel is defined as a container designed to hold gases or liquids at a pressure substantially different from the ambient pressure. These vessels are normally bespoke and designed for specific uses. The most common design standards are ASME VIII, PD5500, and EN 13445. Using the appropriate standard, the pressure vessel engineer will design and specify for:

- Pressure envelope
- Nozzle loads
- Fatigue loads

Modifications to existing pressure vessels may be required for:

- Addition of or modification to nozzles.
- Recertification.
- Vessel repairs.

Common pressure vessels types include:

- Separators.
- Heat exchangers.
- Filters.

Cranes and Lifting Equipment

Specification and modification to cranes and lifting equipment is a specialised area and is normally carried out by the engineering department of companies specialised in lifting.

Proprietary/Specialist Equipment

The mechanical package engineer will also be involved in the management of specialist equipment packages such as lifeboats and firefighting equipment.

Studies

The mechanical discipline often carries out the following studies ahead of enquiry in the early phases of projects to allow for selection of the optimum equipment to meet project needs:

- Best available technology assessment.
- System/equipment evaluation.
- Multi-decision criteria assessment.

Best available technology assessment (BAT)

In these evaluations, the mechanical engineer will evaluate all the available technologies to determine which is best and within acceptable cost limits for the intended application.

Systems/equipment evaluation

Systems and equipment packages will be evaluated against many criteria including performance reliability, supplier track record, cost, and maintainability. These features are often conflicting and one might use "multi-decision criteria analysis" techniques to select the preferred equipment and supplier.

Multi-criteria decision analysis

This technique evaluates multiple conflicting criteria in decision making. Rather than relying only on engineering intuition, it frames the selection criteria in a structured manner and works towards an agreed consensus based on an iterative review of the most important requirements.

Routes to Becoming a Mechanical Engineer

Today, most mechanical engineers have a degree in mechanical engineering with experience in their speciality arising from time working with an equipment supplier or the main contractor or operator through a graduate recruitment route.

Mechanical Summary

Most offshore equipment is of a mechanical type, be it a pump, vessel, or equipment package. Equipment packages can be considered stand-alone items that are then integrated into the overall platform production process via agreed connection interfaces. Attention to detail, including supplier documents and inspection during manufacture, ensures that what has been specified is delivered.

PETER F CRANSTON

THE PROJECT ENGINEER'S TOOLKIT•181

CHAPTER 22

Structural

This chapter will take you through the main platform components designed by the structural design process and some of the challenges you might experience.

I recommend that one of the first documents to seek out and study when you join a project are the structural general arrangement, (GA), plans and elevations that allow you to orientate yourself with the layout and locations of the key parts of the facility.

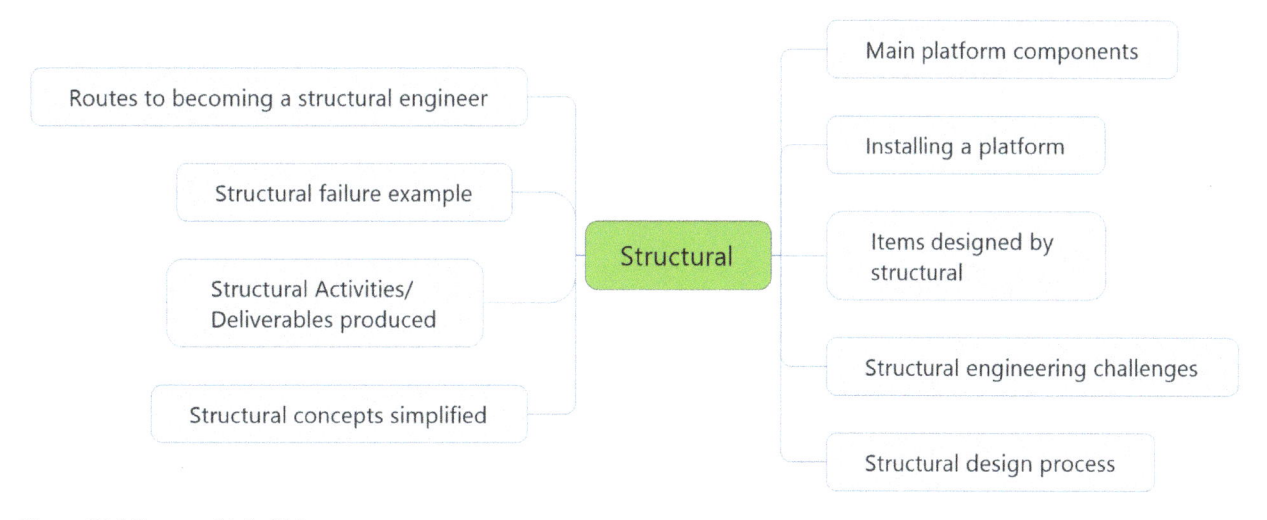

Figure 21.1 Structural Mind Map.

Main Platform Components

An Oil and Gas platform is composed of the main components shown in Figure 22.2 below, although there will be individual variations. When the offshore Oil and Gas industry initiated marine crane technology, lifting capacity was more limited and platform topsides were constructed of many modules with long duration man-hour intensive hook ups. These days it is not uncommon to have lifts of integrated decks in excess of 10,000 tonnes with minimal hook-up and time to production.

More recent marine lifting vessels have pushed the boundaries of single lifts beyond the 20,000 tonne mark, with recent decommissioning of the Brent Delta lift coming in at circa 24,000 tonne single lift using the purpose-built Allseas' single-lift installation/decommissioning and pipelay vessel, "Pioneering Spirit".

CHAPTER 22 STRUCTURAL

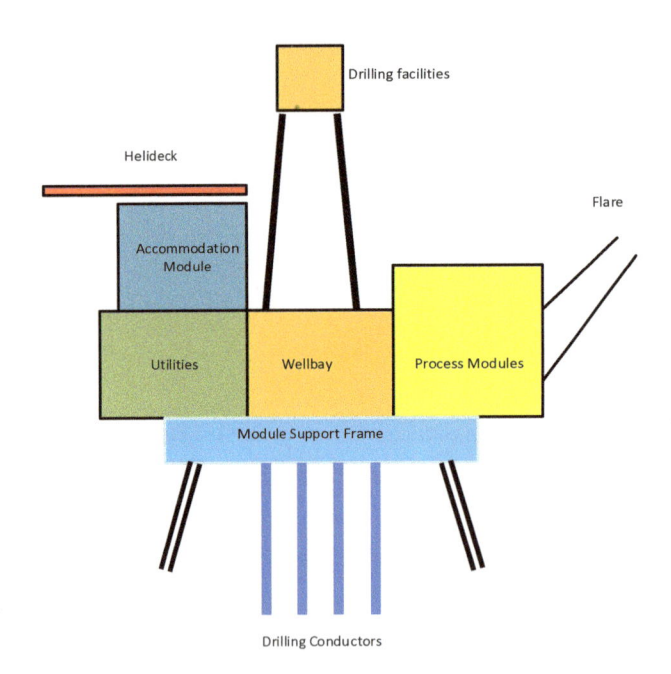

Figure 22.2 Platform topsides simplified block diagram.

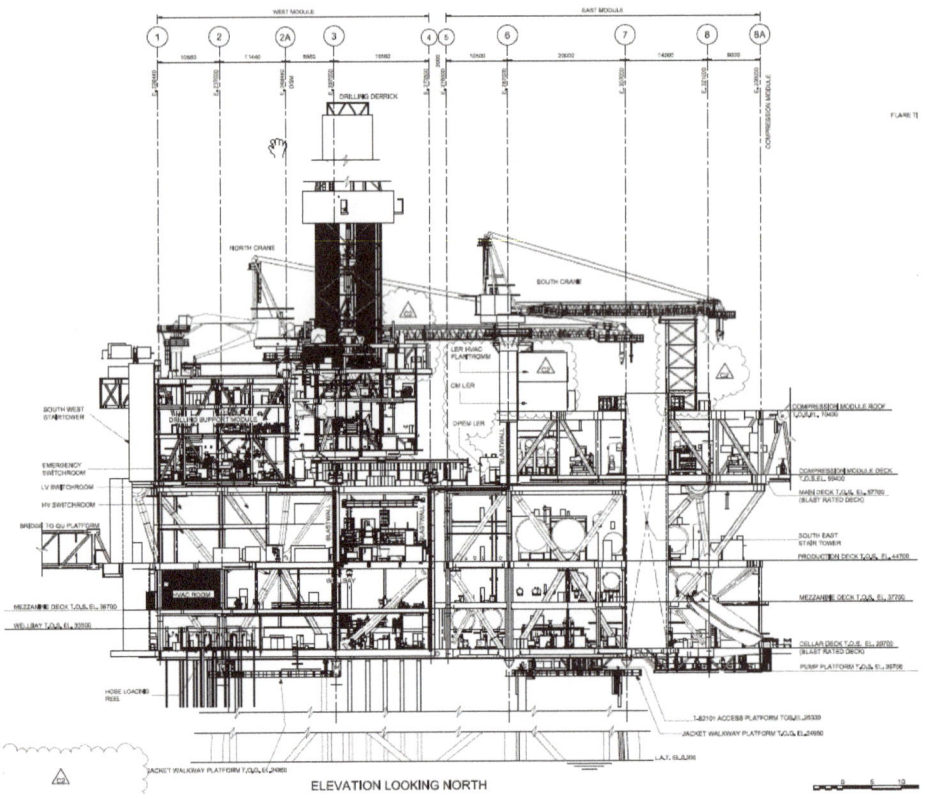

Reproduced by permission of BP Plc.

Figure 22.3 Typical platform elevation.

PETER F CRANSTON

Installing an Offshore platform

Let's now build a platform from the seabed up! Construction starts with the installation of the jacket, the structure that sits on the seabed and supports the topsides structure and modules. This is sometimes installed on a pre-drilled manifold to reduce time to production. The jacket is towed out to sea by tugs and can be lifted vertically into position, skidded off the edge of a transport barge, or floated into position and up-righted by selective de-ballasting.

Once temporarily located in position on the seabed, the jacket is secured using preinstalled pile guides on the main legs, or alternatively, if the seawater depth is not significant, internally through the jacket legs. There are other alternative installation types, notably Gravity Base Structures (GBS), Tension Leg Platforms (TLP), Spar Platforms, and Semi Sub to note a few. Now with the jacket in position, the installation of the topsides modules takes place.

Newer platforms usually have a large, partially commissioned, and integrated deck that is lifted on top of the jacket that are temporarily secured into place with stab-in guides on top of the jacket legs, providing horizontal restraint to allow weld-out of the topsides for the permanent conditions.

After the integrated deck is in place, the accommodation and drilling modules can be lifted into place followed by the flare tower.

Heavy lifts

Heavy lifts are impressive events with the crane barges dwarfing the facility. One hundred percent pre-preparation is required because if you burn through the sea fastenings securing the module to the transportation barge, there are no second chances. During a heavy lift I was involved in for a 2000 tonne module, I asked the heavy lifting engineer what they would do if the module did not fit. He immediately replied that they would lay it down on the seabed because it could not be returned to the barge. Food for thought when preparing for your heavy lift!

Reproduced by permission of BP Plc.

Figure 22.4 Heavy lift topsides installation

There are many detailed variations to this basic sequence, but all are essentially similar.

CHAPTER 22 STRUCTURAL

Items Designed by Structural

Primary structures

The following items are what are known as primary structures (details on what is defined as primary can be found in EEMUA 158, the construction specification for fixed offshore structures):

- Jacket.
- Integrated deck.
- Process modules.

A primary structure is critical to the global integrity of the asset and failure of primary structural members would risk danger of collapse of the asset. For these reasons, the applicable design codes and materials and fabrication requirements are more onerous, requiring all work to be fully checked at all stages.

Secondary structures

Secondary structures are load-bearing structures that are not critical to the integrity of the asset. Local damage would occur if these structures failed, but the asset would remain intact (details on what is defined as secondary can be found in EEMUA 197, the specification for the fabrication of non-primary structural steelwork for offshore installations). Secondary structures include:

- Deck stringers.
- Deck plates.
- Access platforms.
- Stair towers.
- Equipment supports.
- Protection frames.
- Piperacks.
- Pipe supports.
- Passive fire protection.

Tertiary structures

Tertiary structures are low and non-load bearing structures that include:

- Handrails.
- Ladders.
- Gratings.
- Drip trays.
- Drain boxes.
- Gutters and penetration sleeves.

Installation aids and mechanical handling

Structural engineering is also responsible for the design and specification of rigging and lifting frames and for the installation and removal of permanent and temporary equipment, such as lifting frames, spreader bars, pad eyes, and runway beams.

Multi-discipline supports

While equipment installed by other disciplines requires supports, the supports themselves are normally designed by the structural teams.

Structural Engineering Challenges

Weight

Offshore installation jackets have "not-to-exceed" weight limitations that are determined during the initial design. A weight control engineer will be tasked with managing this weight budget through the course of the engineering design phase. They will record the weights of all the proposed structural steelwork, piping, and equipment. The design must satisfy both the imposed design codes and the imposed weight limit.

The weight estimate is managed in a similar manner to a financial budget, with a large contingency early in the project that is progressively reduced as the design, fabrication, and construction progress. The final part of the weight control is actual weighing of equipment and modules before they are finally installed.

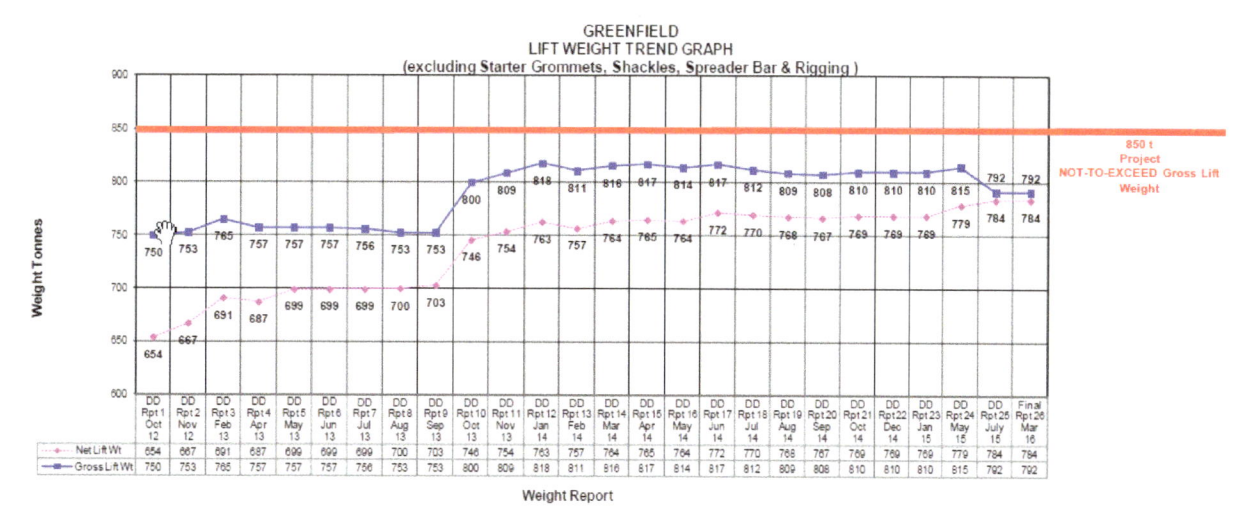

Figure 22.5 Example weight trend graph.

Space

Space is always at a premium offshore. For many brownfield projects, the only way to obtain space is to destruct redundant equipment or to cantilever equipment off the existing structure.

Constructability

Designing for a greenfield build in a construction yard gives opportunity to modify and identify the optimum lifting and installation methodologies, and in some respects can be considered to give more flexibility to the engineering team over brownfield installations. For a brownfield project, installing onto a live facility from a marine vessel, the design must take into account how the existing asset will accommodate installation guides and protection requirements to ensure that the equipment/module is guided safely into place whilst minimising the impact to the offshore production.

"Tail End Charlies"

This is not really a structural issue per se but is still worth mentioning. Structural engineering is usually last in the detail design process but first in terms of required fabrications. The module or support framing is required to allow equipment and piping to be fitted in, putting a significant emphasis on ensuring that the design information is conservative, allowing for any changes during the design process.

The Structural Design Process

The structural design process starts with a requirement advised by the client as a stand-alone structural workscope, or more commonly as part of an overall project or modification to existing facilities.

Greenfield projects

In a greenfield project, the structural design will be initiated in the study phase moving through FEED and detail design, with greater detail and definition determined at each stage as layouts and equipment weights become more defined. Detail will be done in a 3-D Computer Aided Design (CAD) system involving multiple disciplines. The output will be 2-D fabrication drawings issued to the fabricator.

Brownfield projects

For brownfield projects, the process is different. Typically, scopes cover installation of additional modules, items of equipment, or modifications. If you are fortunate, there will be a 3-D model available for initial evaluation and sizing, potentially with specialist laser surveys of the area obtained from legacy engineering modifications.

Irrespective of the data available, the designer/engineer should complete an offshore survey to photograph and measure the worksite in order to identify differences from the 3-D model. They will then return onshore and complete the design. A further verification survey will usually be carried out with the final design to ensure there aren't any clashes or installation problems.

Tip: Never fully rely on a 3-D model for brownfield scope. Always, always carry out an offshore survey as there will **always** be differences from the model.

Whilst the survey engineer will use a tape measure and camera, there are specialist survey companies who will carry out very accurate point-to-point surveys using laser equipment, which is vital when measuring critical dimensions such as module installations with guide systems. There are also point cloud surveys in which a scanning laser is placed at locations in the installation to build an accurate 3-D picture of the module and equipment.

Structural Concepts Simplified

Unless you have a background in structural or mechanical engineering, you are unlikely to be familiar with the key structural concepts:

Loads

These are the input data for structural, involving a range of loads to allow structural engineers to complete their design, such as:

- Dead loads (self-weight of the equipment/module itself with its contents).
- Live loads (installation lifting loads).
- Environmental (including wind and wave).
- Thermal loads.
- Blast overpressure loads.

Structural philosophy

In the design of any structure, it is important to be clear about the design philosophy and keeping the solutions simple. However, there are many types of solutions to most problems and the structural engineers will consider several methods:

- Braced Structure.
- Cantilevered Structure.
- Propped Cantilevers.
- Moment Frames.
- Trussed Frames.

Design codes and design verification

There are a range of design codes for offshore structures, the common ones being:

- AISC.
- Eurocodes.
- API.

There are a number of types of codes from Working Stress Design (WSD) to Limit State Design (LSD). The WSD applies simple factors reducing the allowable stress within an element being designed while LSD applies factors to the applied loads depending on the type and application of the loads to define a maximum member capacity. Codes are regularly updated and most recently codes have started to move to Load Resistance Factor Design (LRFD), which utilises separate factors for applied loads and material resistance factors. The development of these codes is largely as a result of better definitions regarding the application and use of the materials in order to ensure greater efficiencies of design.

Utilisation

You will often hear your structural colleagues talking about utilisation of a structural member, which is often quoted as a percentage or a figure between 0 and 1 or 0% and 100%. A 100% utilised section can be considered to be loaded to the maximum allowable stress level for that particular material and application per the appropriate design code.

Due to either physical failure of existing structures a design issue on a proposed structure, a member can have a reported utilisation greater than 100% after initial analysis .However, further refinement of the calculations or alternatively strengthening the elements can be undertaken to allow a project to continue or resolve concerns over individual elements.

Allowable stresses

The engineers then complete calculations, usually using a mixture of software and manual methods to calculate the stresses in a particular design. Stress results when a material is loaded with a force. Depending on the design code, the proposed material will have a maximum allowable stress. If the calculated stress is too high, the engineer might increase the area of the section of material used or change the design geometry. The symbol for stress is σ, (sigma), and is measured in N/mm^2.

Allowable strains

Strain is a term used to measure the deformation or extension of a body subjected to a force or set of forces. The symbol for strain is ε, (epsilon), and is dimensionless as it is the ratio of the original unloaded dimension over the loaded dimension. Strain is used to determine deflections in loaded structures. Some equipment, such as gas turbine generators and pumps, are critical to the amount of deflection in their foundation supports which may cause alignment and vibration problems if of a high value.

There is always a trade-off in structures as increasing section thickness may reduce stresses and strains but will cost more and will increase the overall weight of the topsides.

Structural Activities and Documents Produced

Structural integrity model holder

Engineering contractors normally hold the structural integrity models for offshore assets on behalf of Oil & Gas operators. These are computer models, reflecting the as-built structural design and containing utilisation information of all the primary, and in some instances some secondary, structural members.

Structural integrity models are used when damage or corrosion is identified or when additional modules or equipment are proposed. These models allow structural engineers to evaluate the integrity of the structure and consider any modifications or repairs that must be conducted. The models are updated annually by most operators as part of their integrity programme, along with the platform weight control reports.

The models may also have interim updates associated with specific project weight changes in which large projects change the weight of an asset. An assessment at FEED should be conducted to demonstrate the impact of the proposed modification to the asset and ensure structural integrity is maintained.

Structural analysis reports

Engineering contractors will carry out analyses usually using commercially available software, to determine loads and utilisations in proposed designs prior to fabrication. They will provide these reports to the structural model holder to carry out 'model integrity checks'.

Structural engineers will also prepare detailed internal design reports detailing the input data, design methodology, applicable codes and standards, and analysis results.

Verification

It is the responsibility of the engineering contractor to ensure that the design is code-compliant. The engineering contractor will produce a design report which contains the overall design plus the supporting code compliant calculations, which will be then submitted to an independent verifier such as Lloyds, DNV, or Bureau Veritas, for independent engineers to check and confirm that the design is code-compliant. This process provides assurance that the design is robust. Since the process can be lengthy, the reviews should be scheduled in a timely manner to be sufficiently complete before any fabrication subcontracts are placed. It is not uncommon for verification to be ongoing with onshore fabrication nearing completion, resulting in increased pressure on the project team.

Other calculations

Structural engineers will also carry out calculations by hand or assisted by Mathcad™ on simpler elements to check code compliance.

Structural fabrication drawings and specifications

Structural engineers assisted by designers produce fabrication drawings and specifications which are then issued to fabricators for fabrication of modules, structures and supports.

Supply chain will issue these for enquiry. Once proposals have been received, the structural team will review them for technical compliance. If non-compliant, supply chain might work with the client and fabricator to agree on any required deviations to allow fabrication to be carried out.

Weight control reports

Detailed weight control reports will be issued regularly, collating inputs from other disciplines and advising any weight issues and options for resolution. On large weight sensitive projects, it is key to have a defined weight control strategy and regular weight review board meetings to ensure control of weight changes are managed.

Criticality rating assessments

Depending on the end use of the structural design, a criticality rating will be assessed. The more critical the structure, the more stringent the materials specifications, weld quality, and other inspection.

2-D and 3-D drawings

The structural discipline generally works in a controlled 3-D model and will output 2-D fabrication drawings containing full fabrication details, including bills of materials.

Offshore surveys

Detailed offshore survey reports will be generated, containing sketches, key dimensions and supporting photographs.

Example of Structural Failure: The Alexander Kielland

In March 1980, the semi-submersible accommodation rig, the Alexander Kielland, capsized, killing 123 people[Ref9]. An investigative report concluded that the rig collapsed due to a fatigue crack in one of its six main bracings. This was traced to a small 6mm fillet weld.

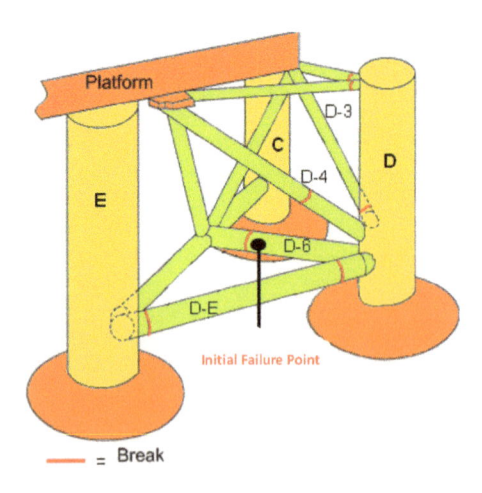

Figure 22.6 Alexander Kielland structural failure.

Like most incidents, there were a number of contributing factors. A poor weld profile, cold cracks in weld, and cyclical stresses due to offshore environmental loads. A crack in a seemingly non-critical area associated with a hydrophone attachment propagated and caused a main structural member (D6 above) to fail. Due to a lack of design redundancy, other main members then sequentially failed, resulting in the rig capsizing. The original structural design was a significant contributor to the failure.

Routes to Becoming a Structural Engineer

The normal route to becoming a structural engineer is via a graduate degree in either structural or civil engineering and then by gaining experience and seniority in an engineering contractor or specialist consultancy. An initial professional development stage is normally undertaken by graduates, depending on individual aspirations but typically considered to be around a four year period.

The other route is to start as a structural designer and, after gaining experience as a designer producing structural drawings, obtaining a structural degree and transitioning to be an engineer.

Structural Summary

Structures support everything offshore. From the complete topside modules to individual pipes and equipment items. Given the hostile environment offshore, the large structural loads imposed, and the consequences of structure failure, it is vital that primary and secondary structures are fully code-compliant in design and that rigorous QA/QC supports fabrication and subsequent maintenance inspection.

CHAPTER 23

Instruments and Control

Offshore Oil and Gas control and instrumentation systems are immensely complex and can initially intimidate the non-specialist. This chapter will explain the general architecture of the main systems and cover the main field equipment that connects to these systems.

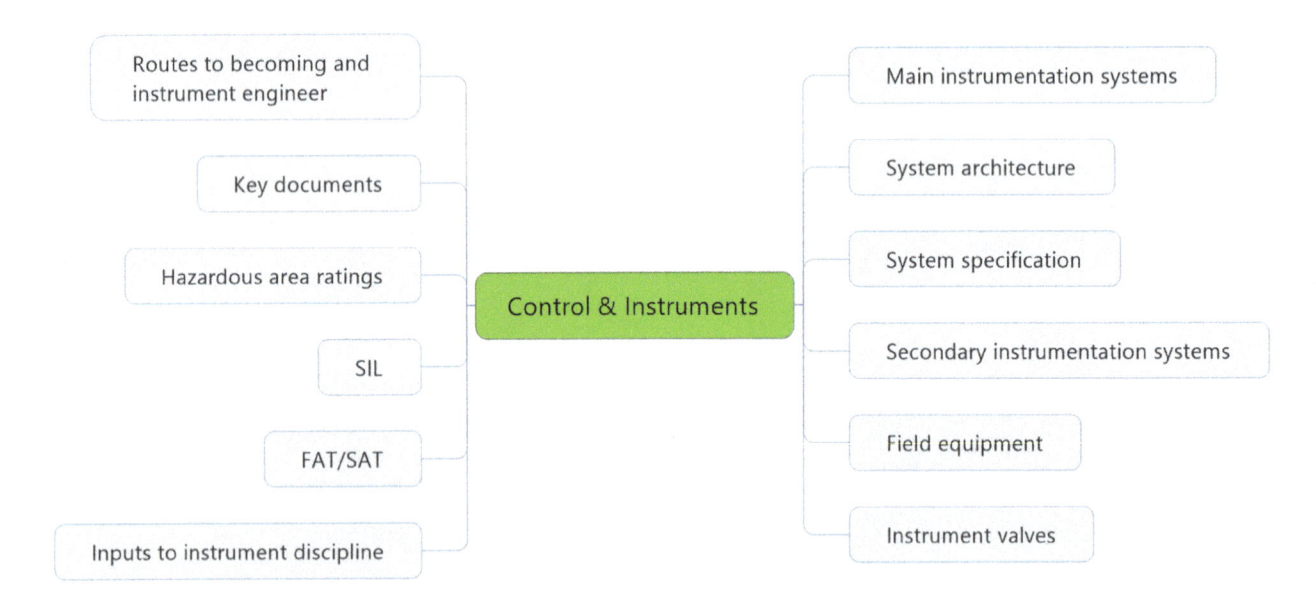

Figure 23.1 Instruments and control Mind Map.

The simplest way to think of the control and instrumentation systems for offshore facilities is to liken it to the brain, nervous system, and senses of a human. We will first look at Instrumentation Systems, and then Field Instruments.

Main Instrumentation Systems

There are three main systems which control an offshore facility:

- Distributed Control system. (DCS)
- Fire and Gas system. (F&G)
- Emergency Shutdown system.

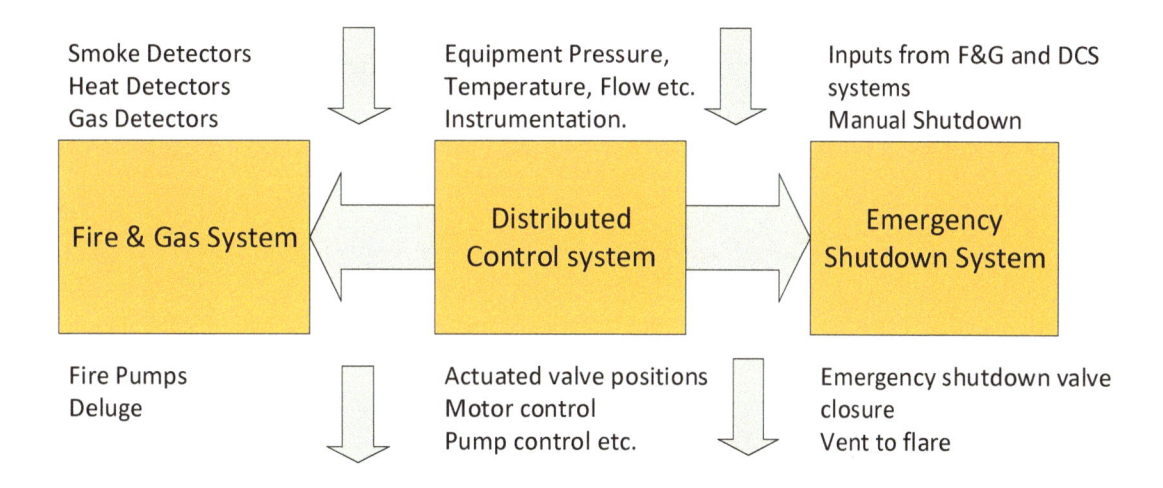

Figure 23.2 Simplified control systems.

Distributed control system

This is the system that controls the production process facility. Platform operations personnel love steady state operations, however, there will be times when things will change due to planned activities such as bringing on new wells, or issues requiring shutdown of parts of the system. The system is generally automated with manual operator intervention when required

Fire and gas system

The fire and gas system, as the name suggests, allows for early warning of any fire or gas release issues which, if left alone, could quickly escalate into a serious event. The fire and gas system consists of gas, heat, and flame detection instruments at key locations all over the facility. Upon detection of gas, heat, or flame, the system will alarm in the control room and allow either automatic or manual actions to be taken. Manual Alarm Call-points (MACs) located in the plant and on the equipment, allow for manual initiation of the fire and gas system on detection of a problem.

Emergency shutdown system

In a particular set of circumstance, the emergency shutdown (ESD) system will automatically shutdown part or all of the facility. Depending on the severity of the event, shutdown can range from shutting down part of the process to shutting down the complete field and blowdown of the hydrocarbon inventory.

Piper Alpha

The importance of robust ESD systems cannot be underestimated. In 1988, when platform emergency shutdown systems were less sophisticated, 167 lives were lost on the Piper Alpha when a gas release ignited. On the Piper Alpha, the initial fire was continually fueled by hydrocarbon inventory from pipelines feeding into the facility. As a result of this tragedy and the resulting Cullen enquiry into the Piper Alpha disaster [Ref10], more robust ESD systems were stipulated and retrofitted to all United Kingdom Continental Shelf UKCS offshore facilities.

CHAPTER 23 INSTRUMENTS AND CONTROL

ESD blow down

The basic philosophy with a modern emergency shutdown system is to close in the wells and oil/gas risers feeding the platform and to vent the platform gas inventory to the flare within a short period. This ensures that the incident does not escalate and allows for subsequent escape and evacuation of personnel, if required.

Platform systems evolution

The earlier first- and second-generation platforms there were typically three separate systems often provided by different manufacturers with complex interfaces:

- Fire and Gas System.
- Emergency Shutdown System.
- Distributed Control System.

More recent facilities, however, use Integrated Control and Safety Systems (ICSS) that combine all of the above systems into a single system supplied by one manufacturer.

System Architecture

We will use the Distributed Control System (DCS) to explain the configuration and operation of a typical system. However, the Fire & Gas and Emergency Shutdown systems are similar in concept.

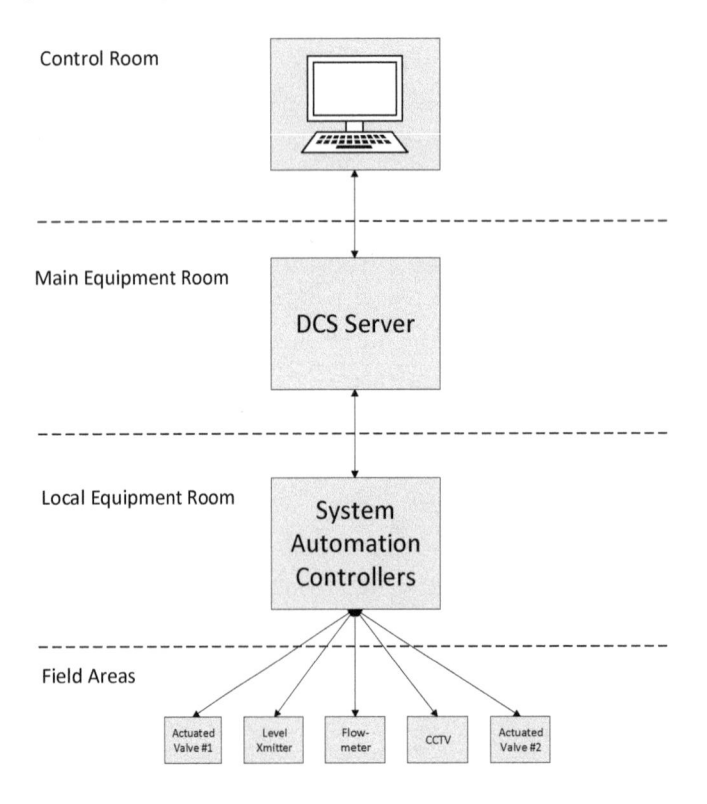

Figure 23.3 Simplified DCS architecture including physical locations.

Original control systems on first-generation facilities used hard-wired, individual controllers, pneumatic controllers and relay logic systems. Modern DCS system use industrial PCs and networks with industry standard communication protocols.

To ensure maximum reliability, many parts of the system are duplicated. This redundancy and 'hot standby' ensures that even if part of the system fails, the overall system remains operational.

System Specification

Given the complexity and individual requirements of offshore facility control systems, each DCS system is bespoke using a supplier's system that is configured and programmed for the client's specific requirements.

Cause and effects

The key inputs to the design of the ESD and F&G systems includes overall system design philosophies and cause and effects. Simply put, if a series of events (causes) happen, a series of outputs (effects) will occur.

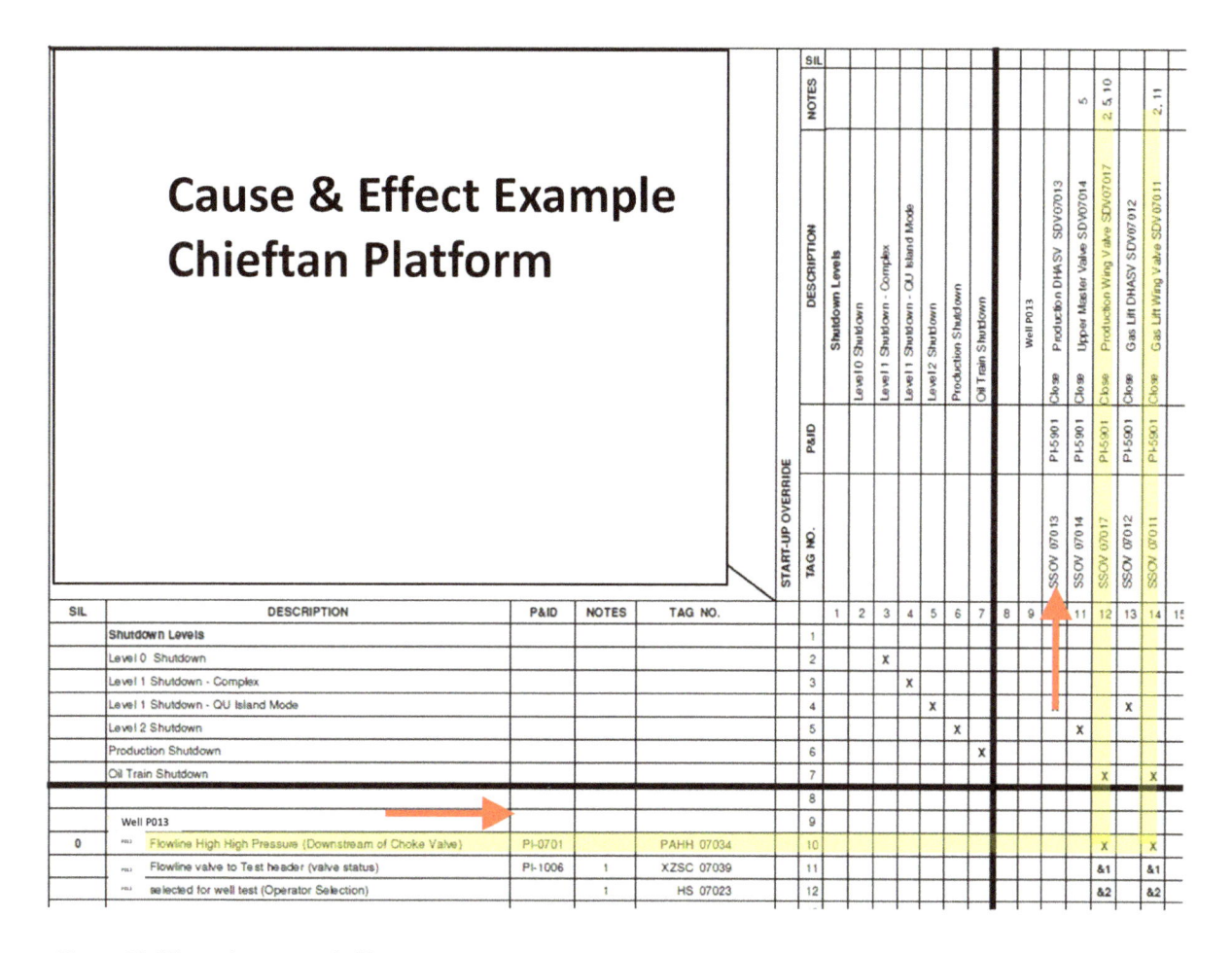

Figure 23.4 Example cause and effects.

CHAPTER 23 INSTRUMENTS AND CONTROL

The cause and effects detail the required control output (effect) based on a certain set of instrumentation or operator inputs (causes) and originates from the process discipline. For the above highlighted example, follow the high-high pressure signal horizontally to the intersection marked "X" and follow vertically to the automatic valves closure actions (effects).

Input/Output types

There are two types of input/output (I/O) to DCS systems:

- Digital: On or off indication/instruction.
- Analogue: variable signal between set levels.

You will often hear instrument systems engineers talking about I/O quantity. This is simply the number of digital and analogue inputs and outputs a system needs. When specifying a system, the engineer will typically add 20% additional I/O capacity to allow for future expansion and modifications to the system. A typical platform DCS system may have many thousand I/Os.

Graphical interface/ human machine interface

Graphic or human machine interfaces are the touch screens and keyboards that platform control room operators use to monitor and control the process plant.

Part of the scope of supply by system providers is to incorporate the computer graphics that schematically show the process configuration and status of equipment and process. It is important that these graphics are clear and easy to use, especially in an emergency event.

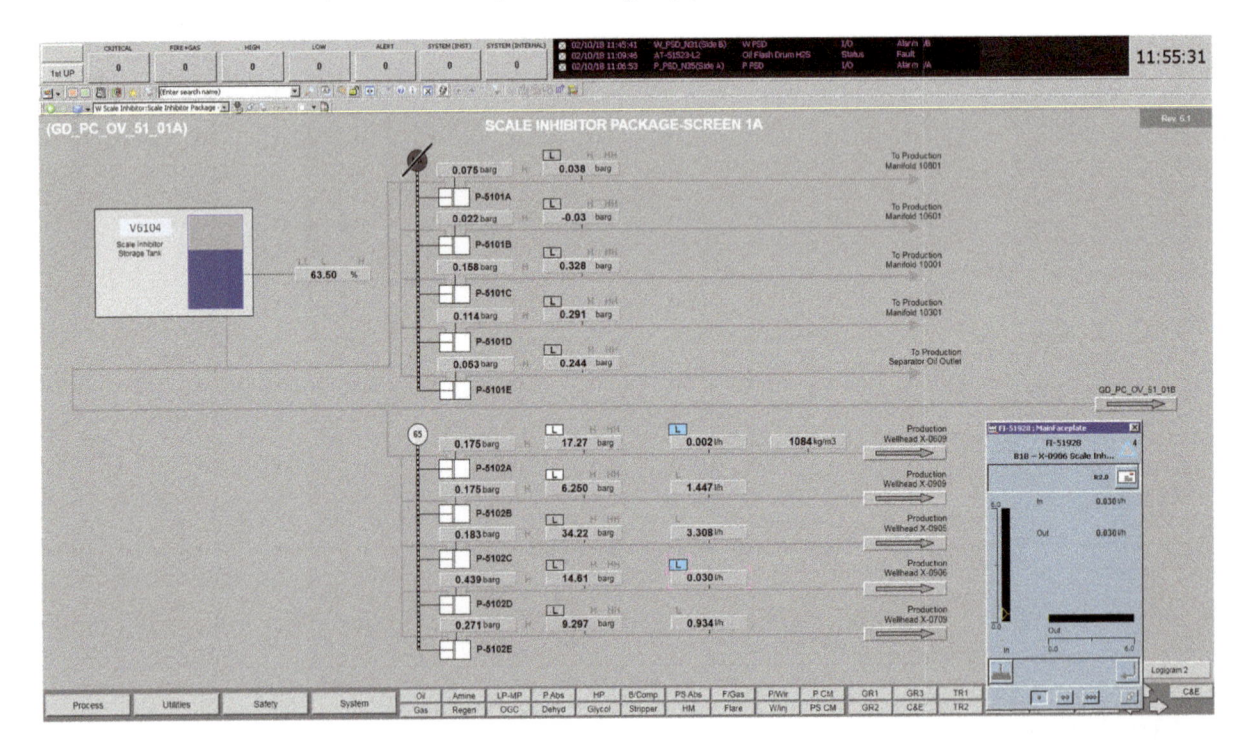

Figure 23.5 Typical control room screen graphic.

PETER F CRANSTON

Communication with onshore team

Control and monitoring does not stop offshore. Using the various telecommunication systems, the offshore graphical interfaces are replicated onshore, allowing rapid support from the onshore operations engineering team. Data from instrumentation is also recorded and trended to allow analysis and resolution of problems.

Equipment obsolescence

Obsolescence of equipment and components is a major issue for facility control systems resulting in requirements to upgrade and change-out control systems. These major upgrades must be done with minimum or nil shutdown and can be likened to changing out a person's brain and nervous system whilst they are still alive.

The "steady hand"

We cannot finish platform control systems without mentioning the "Steady Hand". Working on live DCS and ESD system by the vendor specialists requires a steady hand when working in live control panels. An inadvertent slip of the screwdriver or disturbance of cabling can result in expensive, multi-million dollar unplanned shutdowns which may take many hours to restore platform production.

Secondary Instrumentation Systems

Other secondary systems that you will find on an offshore facility are the:

Telecommunication systems

Telecommunication systems provide voice, video, and data links with onshore and other offshore installations plus personnel announcements (PA) and general alarms (GA) for the platform.

Corrosion monitoring

These passive devices can be connected to pipework and, using ultrasonics, measure and report wall thickness changes.

Sand monitoring

These passive devices can be connected to pipework and, using ultrasonics/acoustics, measure and report sand rates.

Condition monitoring

These systems are normally connected to major pieces of the plant, such as pumps, compressors and gas turbines, and can measure and report vibration, speeds, temperature, pressure, and oil analysis. This information can be used to predict problems and allow planned maintenance.

Metering systems including fiscal systems

From the Oil and Gas accountants and taxation authority viewpoint, this is the most important system on the facility. Fiscal metering systems are installed in the process at the point of export either to a pipeline or to an offloading tanker. These highly accurate and regularly calibrated systems measure the flowrate and composition of the exported product.

CHAPTER 23 INSTRUMENTS AND CONTROL

Field Equipment

These are the pieces of equipment located on the facility which either passively sense process and equipment parameters or actively control equipment.

The following list and mind map detail the main types of instruments and the sub-types.

Main categories of field equipment

- Pressure.
- Flow.
- Temperature.
- Level.
- Fire and gas detection.

Figure 23.6 Field instruments.

Space does not permit a full explanation of the principles of each element, but more detail, if required, can be found by a general internet search.

PETER F CRANSTON

Actuated Instrument Valves

Valves that are automatically or remotely actuated are specified and procured by the instrument discipline whilst manual valves are procured by piping. Actuated valves can range in size from less than 1" up to 42" for large subsea pipelines, with every size in between.

There are three main types of instrument valves:

- Relief valves.
- Shutoff valves.
- Control valves.

Relief valves

Relief valves are calibrated valves which open to relieve pressure once a predetermined value is reached in order to protect pipework and vessels from damage.

Consider the following example of a vessel with a damaged relief valve. There is a process problem in which gas arrives in the vessel at high pressure, causing the vessel to become pressurized above its maximum working pressure. The relief valve fails to operate and the pressure continues to rise until the vessel ruptures with a large uncontrolled gas leak. A working relief valve would vent this pressure at a predetermined value well below the vessel failure pressure and route this gas to the flare system, where it can be safely vented.

Shutoff valves

Shutoff valves perform an on/off operation and are typically ball or gate valves that are required to provide consistent leak-tight performance.

Control valves

Control valves are capable of continuous partial openings to control the flow of a fluid or gas through the plant.

Other control inputs/outputs

Other controlled systems can include motor driven pumps and gas turbines, in which a range of digital and analogue signals are transmitted to and from equipment, allowing for controlled operation.

Instrument discipline responsibilities

Instruments are responsible for specifications, vendor liaisons, and final FAT acceptance on instrument systems and field instruments.

Inputs to Instrument Engineering

The main inputs to allow instruments to specify systems and equipment are:

- Process AFD P&IDs.

CHAPTER 23 INSTRUMENTS AND CONTROL

- Process data sheets.
- Fire and gas layout drawings.
- Cause and effects.

They will then prepare scopes of work for each of the required systems (greenfield projects) or systems modifications (brownfield projects). Once the supplier order has been awarded, they will liaise with the vendor to ensure that the requirements are being supplied.

Factory Acceptance Testing and Site Acceptance Testing

FAT

A key stage in the procurement process is the factory acceptance test (FAT), which can take many weeks for complex systems. FAT usually involves a computer simulation of the platform connected to the new DCS/ESD system hardware and software to prove the operation for all operating configurations in-line with the cause and effects.

Successful FAT completion is never 100% the first time and will normally require both hardware and software modifications to pass testing.

SAT

Once installed offshore, a Site Acceptance Test (SAT) will be carried out to ensure that the system works as planned offshore. Again, minor modifications may be required.

Field instruments specification and testing

Field instruments are specified by the instrument discipline, and in many cases are sourced directly from suppliers' catalogues. There may be additional bespoke requirements regarding materials or testing, and certification to comply with client standards. However, there is a drive towards standardisation that will ultimately reduce overall engineering and procurement costs.

Mechanical and electrical systems interface

As instruments are involved with anything that needs to be monitored and controlled, they have a close relationship with the mechanical discipline managing equipment packages and with the electrical discipline procuring electrical equipment and systems.

Safety Integrity Level

Safety integrity level (SIL) is defined as a relative level of risk-reduction provided by a safety function or to specify a target level of risk reduction. In simple terms, SIL is a measurement of performance required for a safety instrumented function (SIF). Four SILs are defined, with SIL 4 being the most dependable and SIL 1 being the least.

Typically, in an offshore environment the highest integrity requirement will be SIL3. The design should aim to minimise high SIL requirements through the use of inherently safe designs because from a project perspective, higher SIL requirements incur greater overall costs. This is covered in more detail in Chapter 25, "Technical Safety".

PETER F CRANSTON

Hazardous Area & Ingress Protection Ratings

You will often hear instrument engineers talking about "Exi, Exd, Exp". The following table summarises what each of these means and where they are required.

Zone	Ex rating	Key features
Non-Hazardous	All	An explosive atmosphere is not expected to be present in quantities as to require special precautions for the construction, installation, and use of equipment.
Zone 2	Exi, Exd, Exe, Exn	Zone 2: An area in which an explosive gas atmosphere is not likely to occur in normal operation and, should it occur, will only exist for a short period of time.
Zone 1	Exi, Exd, Exe	Zone 1: An area in which an explosive gas atmosphere is likely to occur in normal operation.
Zone 0	Exi	Zone 0: An area in which an explosive gas atmosphere is present continuously or for long periods of time.

Table 22.7 Zone/Ex ratings.

IP ratings

The IP Code; International Protection Marking IEC standard 60529, sometimes interpreted as Ingress Protection Marking, classifies and rates the degree of protection provided against intrusion (body parts such as hands and fingers), dust, accidental contact, and water by mechanical casings and electrical enclosures:

- First digit: Solid particle protection.
- Second digit: Liquid ingress protection.

Example: IP67

- First Digit 6: Dust Tight, no ingress of dust, and complete protection.
- Second Digit 7: Ingress of water in harmful quantity shall not be possible when the enclosure is immersed in water under defined conditions of pressure and time (up to 1 meter).

Note: The higher the IP coding numbers the higher the protection.

CHAPTER 23 INSTRUMENTS AND CONTROL

Intrinsically Safe Systems

Intrinsic safety (IS) is a protection technique for safe operation of electrical/instrumentation equipment in hazardous areas by limiting the energy (electrical and thermal) available for ignition. In signal and control circuits that can operate with low currents and voltages, the intrinsic safety approach simplifies circuits and reduces installation cost over other protection methods. A device termed intrinsically safe is designed to be incapable of producing sufficient heat or sparks that would ignite in an explosive atmosphere, even if the device has experienced deterioration or has been damaged.

High-power circuits such as electric motors or lighting cannot use intrinsic safety methods for protection as they are capable of generating sparks of sufficient energy to ignite a gas leak.

Key Drawings and Documentation

In addition to procurement documentation, Instruments produce drawings that will be included in construction workpacks for the construction and commission teams offshore to refer to as they construct hook-up and commission systems.

Key drawings:

- Layout drawings.
- Interconnection drawings.
- Loop drawings.

Key documents:

- Cause and effects.
- Instrument sizing calculations.
- SIL verification calculations.

Routes to Becoming an Instrument Engineer

The three main routes to becoming an instrument engineer are:

- A degree in Electrical Engineering/Electronics/Instrumentation followed by a graduate instrument engineering career program with a main contractor.
- Completion of a recognised offshore/site instrument technician apprenticeship and experience supported by further study when transitioning to an onshore role.
- Starting as instrument designer/apprentice and progressing to an engineer via experience and additional qualifications.

PETER F CRANSTON

Instruments Summary

As mentioned earlier, Instruments for the non-specialists may appear intimidating. However, by breaking each system into its functions and components, it will be easier to understand and manage the instruments that disciplines input into your project. Instruments interface with many disciplines, and by understanding how they work and interact, you will be able to assist them with efficiently delivering the project on-time and meeting the cost bottom lines.

# Electrical

Offshore installations are like small towns that require electrical power for almost everything, from the main oil pumps to the TV sets. This chapter will take you through the electrical power generation, distribution, and control of offshore installations and explain the main items of electrical equipment.

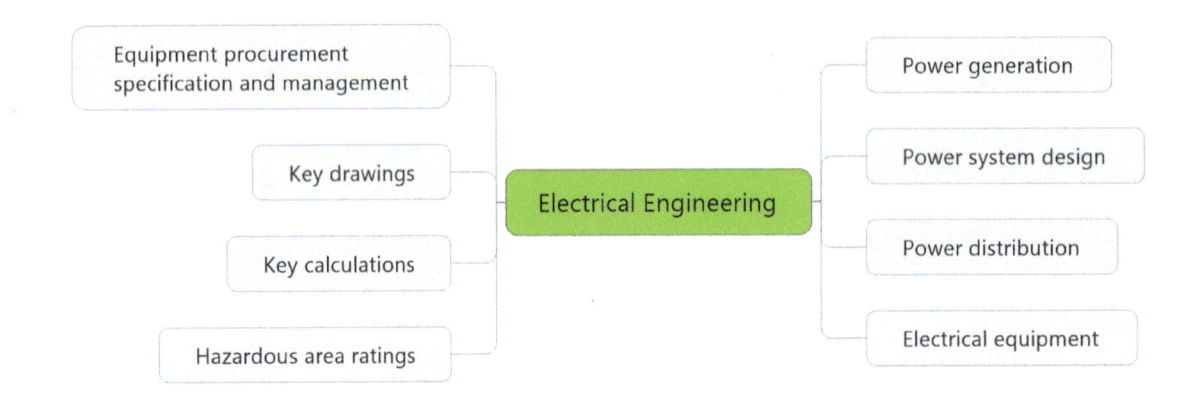

Figure 24.1 Electrical Mind Map.

PETER F CRANSTON

Power Generation

Gas turbines

The main electrical generation on a facility is normally via gas turbines that can be either aero-derivatives or full-sized industrial units.

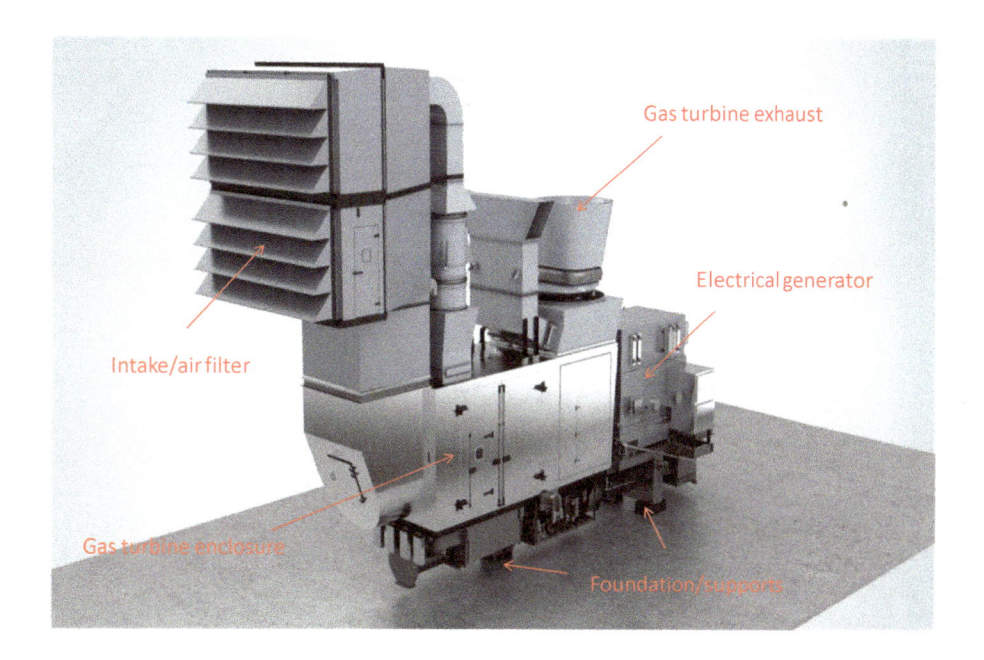

© Siemens AG Communications Power and Gas.

Figure 24.2 Siemens Gas Turbine Generator Unit.

Gas turbines typically started-up using electric motors in which power is provided from diesel generators, which are in turn powered by battery banks.

Gas turbines usually run on either production process fuel gas, or diesel. The diesel capability is provided to start-up the gas turbines before the production process starts producing fuel gas. Gas turbines are synchronised to run together at the same speed, producing power within required frequency and voltage limits.

Diesel power generation and "Black Start"

You may hear the term "black start," which refers to a platform that has no power. Imagine arriving on a platform with no power; you will have to start-up the power from somewhere, and the sequence is as follows:

- Start-up emergency services diesel generator from battery banks or air starters.
- Use electrical power from diesel generator to start up the first gas turbines on diesel.

CHAPTER 24 ELECTRICAL

- Additional gas turbines are then started using power from the first turbine.
- Once the production process is producing fuel gas, change the turbines from diesel to fuel gas.
- Shut down diesel generation.

Load shedding

There will be times when there is not enough power to run the complete platform, such as for operational purposes or during a breakdown. To manage this, a software-based load system continuously monitors the power produced and, if/when the available generated power is reduced, the software will selectively cut power to various systems, prioritising power continuity to the platform safety systems.

Specifying and Designing the Power System

Load list/schedule

The fundamental document used as input to power system design is the load list or load schedule.

ELECTRICAL LOAD SCHEDULE
(Typical example for additional living quarters)

Description	Operating Voltage (V)	Load rating [kW]	Absorbed load [kW]	Full Load Current [A]	Cable Size [mm2]	Load factor [%]	Power factor at L.F.	Continuous [kW]	Continuous [kVAr]	Intermittent [kW]	Intermittent [kVAr]	Stand-by [kW]	Stand-by [kVAr]	Future [kW]	Future [kVAr]
6.6kV/460V ALQ Switchboard Transformer	0	0.00	0.00	0.00	3c x 95	0%	0.00	0.00	0.00	0.00	0.00	0.00	0.00	0.00	0.00
RO Potable Water Maker	440	20.00	20.00	30.87	3c x 6	100%	0.85	20.00	12.39	0.00	0.00	0.00	0.00	0.00	0.00
Calorifier	440	130.00	130.00	170.00	3c x 95	100%	1.00	130.00	0.00	0.00	0.00	0.00	0.00	0.00	0.00
UV Sterlizier	440	6.00	6.00	9.26	4c x 6	100%	0.85	6.00	3.72	0.00	0.00	0.00	0.00	0.00	0.00
440V ALQ Lighting & Small Power Distribution Board	440	35.42	35.42	46.53	4c x 25	100%	1.00	35.42	1.70	0.00	0.00	0.00	0.00	0.00	0.00
400V ALQ Lighting & Small Power Distribution Board	440	15.74	15.74	24.29	4c x 25	100%	0.85	15.74	9.75	0.00	0.00	0.00	0.00	0.00	0.00
Spare	254			0.00		0%		0.00	0.00	0.00	0.00	0.00	0.00	0.00	0.00
AHU Electrical Heater (Via Thyristor Control Panel)	440	60.00	60.00	78.73	3c x 35	100%	1.00	60.00	0.00	0.00	0.00	0.00	0.00	0.00	0.00
AHU Humidifier	440	60.00	60.00	87.00	4c x 35	100%	0.85	60.00	37.18	0.00	0.00	0.00	0.00	0.00	0.00
AHU Supply Fan A Motor	440	11.00	11.00	19.30	3c x 6	100%	0.82	12.09	8.44	0.00	0.00	0.00	0.00	0.00	0.00
AHU Supply Fan B Motor	440	11.00	11.00	19.30	3c x 6	100%	0.82	0.00	0.00	0.00	0.00	12.09	8.44	0.00	0.00
Main Extract Fan A Motor	440	3.00	3.00	5.50	3c x 6	100%	0.81	3.42	2.48	0.00	0.00	0.00	0.00	0.00	0.00
Main Extract Fan B Motor	440	3.00	3.00	5.50	3c x 6	100%	0.81	0.00	0.00	0.00	0.00	3.41	2.47	0.00	0.00
AHU Axial Fan A Motor	440	1.10	1.10	2.10	3c x 6	100%	0.79	1.29	1.00	0.00	0.00	0.00	0.00	0.00	0.00
AHU Axial Fan B Motor	440	1.10	1.10	2.10	3c x 6	100%	0.79	0.00	0.00	0.00	0.00	1.29	1.00	0.00	0.00
AHU Compressor A	440	5.16	5.16	11.60	3c x 6	100%	0.85	5.16	3.20	0.00	0.00	0.00	0.00	0.00	0.00
AHU Compressor B	440	5.16	5.16	11.60	3c x 6	100%	0.85	0.00	0.00	0.00	0.00	5.16	3.20	0.00	0.00

Assumed Maximum Demand — totals: 349.1 | 79.9 | 0.0 | 0.0 | 22.0 | 15.1 | 0.0 | 0.0

Peak of normal running plant load = X1*Cont. load + Y1*Interm. Load	384.03	kW
X1 = 100% Y1 = 50%	87.85	kVAr
Assumed Maximum Demand = 110% of Peak normal load	393.95	kVA
	cos phi	0.97
	34.46 (FLC)	A

kVA: 358.1 | 0.0 | 26.6 | 0.00
cos phi: 0.97 | 0.00 | 0.82 | 0.00

Figure 24.3 Example load list.

The load list identifies all the electrical power loads on the facility and sub-divides them in to the following types:

- Normal: Required for normal running of the production facility.
- Critical: Failure of these supplies will cause temporary or permanent production stoppage.
- Essential: Essential to allow evacuation of the facility.

Note: the exact definition of the terms may vary in specifics by jurisdiction, but all will be similar in meaning.

Load types

Electrical loads are defined as follows:

- Continuous: Demand at all times.
- Intermittent: Typically required 30 to 60% of the time.
- Standby: Typically required 0 to 20% of the time.

The electrical system will must be designed for both the normal consumed load and the peak load.

Power system study

Once the load list has been determined, a power system study can be carried out, consisting of a complex analysis and often carried out by an external specialist. As offshore facilities generate at both high voltages (typically 13.8 kV) and high power (typically 100 MW) for the biggest installations, an incorrectly designed power system can cause imbalances and power surges that might damage equipment. These studies model the entire system and will look at the various configurations and switching scenarios to remove the risk of potentially damaging surge and overload or imbalance situations.

Uninterruptable power supplies

The final parts of the distribution system are the Uninterruptable Power Supplies (UPS) systems that are composed of banks of batteries and invertors that provide power in the event of failure of platform power.

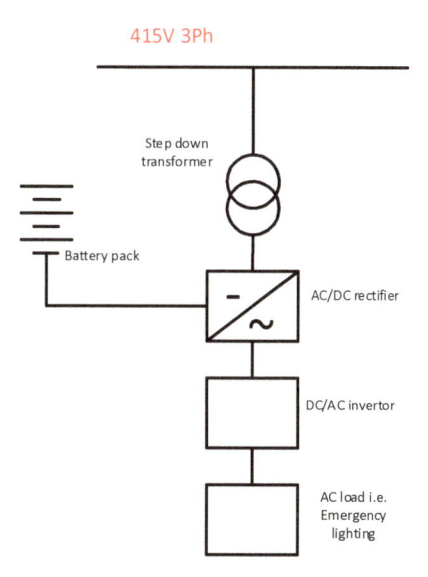

Figure 24.4 Uninterruptible power supply schematic.

CHAPTER 24 ELECTRICAL

UPS systems typically consist of a battery bank and DC to AC invertors that kick in automatically when platform power is lost, providing power to critical systems for a specified period. These protected systems include the ESD, ICS, and other SCS, and are intended to provide power until the emergency is resolved or the facility is evacuated.

Power Distribution

A platform overall power system may at first seem daunting, but by using the single line diagram (SLD), it can be easily broken down and understood.

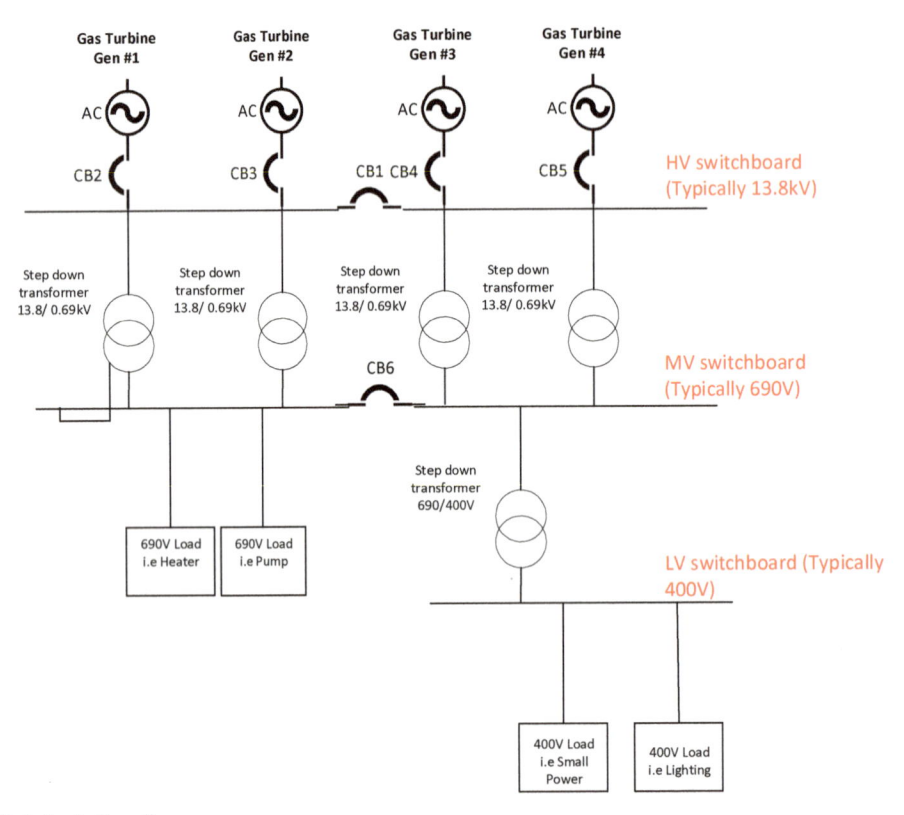

Figure 24.5 Simplified single line diagram.

Power is generated by the gas turbines and distributed to switch rooms located in various parts of the facility where it is distributed further to lower voltage switchboards and ultimately to the final consumers.

Looking at the SLD in the figure above, you will see that there are numerous breakers that allow generators and parts of the system to be selectively connected or isolated, allowing the facility operators flexibility in isolating for maintenance/breakdown work, or for other operational reasons. The SLD also identifies the physical location of the various main switchrooms.

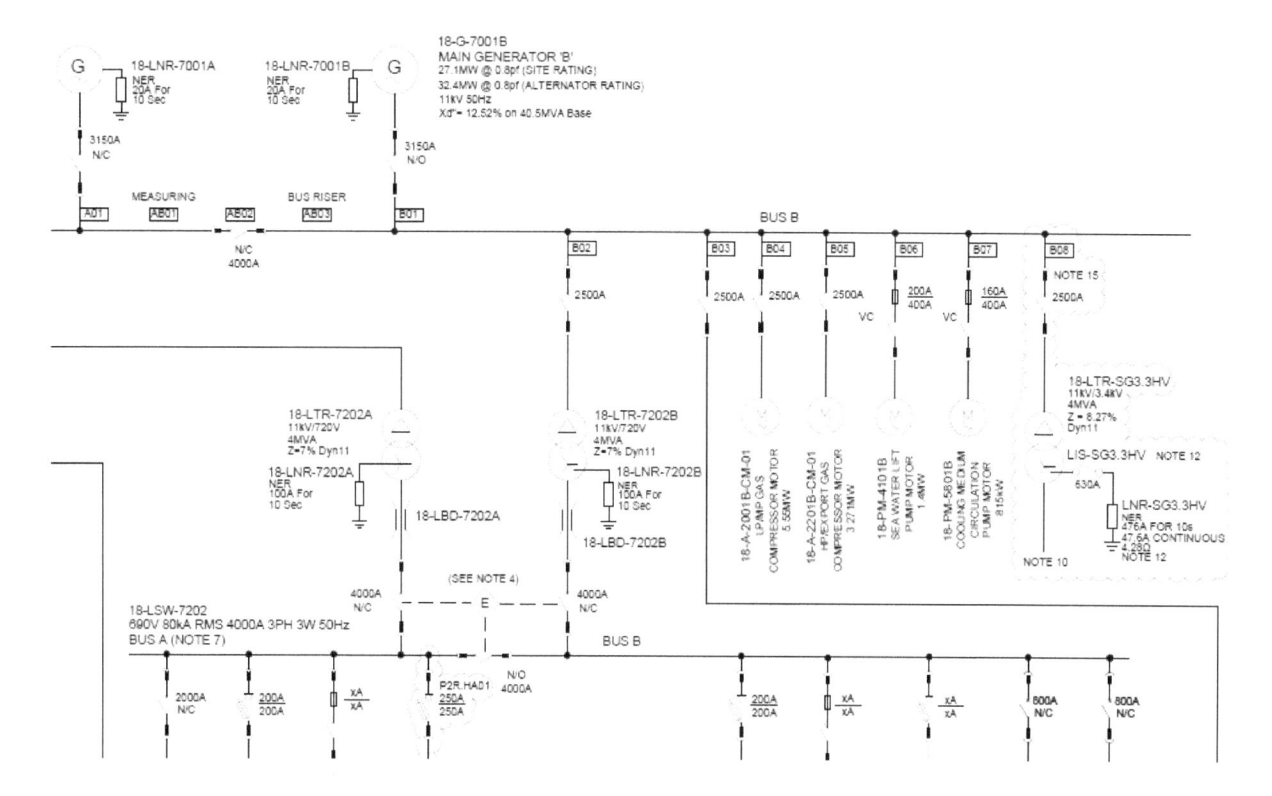

Figure 24.6 Single line diagram excerpt.

Note: The standard electrical engineering definition of high voltage (HV) is 33kV and above, but in offshore installations, 13.8kV is normally referred to as high voltage.

High voltage

The main power is generated at a high voltage to allow efficient transmission to the various switchrooms. High voltage transmission minimises power losses and allows for use of a lower diameter cable, saving on weight and costs. High voltage may be used for main pump drives i.e. main oil line pumps.

Medium voltage

Once in the switchroom, a series of transformers reduces the voltage to typically 690V, which is then sub-distributed to local switchrooms. In some cases, the 690V will be used directly to drive large motors.

Low voltage

The final distribution will be at typically 690/400V 3 phase. Lower voltages can be derived from single-phase connections that provide 200 to 240 V.

Suggested exercise: Using the SLD for your facility, follow the route from the HV main power generation to a local LV switchboard, noting voltages, busbars, breakers, and transformers on route.

Electrical Equipment

Electrical equipment can be split into the following main areas:

- Distribution boards.
- Motor control centres.
- Motors.
- Voltage transformers.
- Heaters.
- Cables.

Distribution boards and switchboards

Distribution boards are main equipment items of the installation electrical supply system that divide electrical power feeds into subsidiary circuits and provide protection for each circuit. In a switchboard, each outlet is served via a cubicle which can be individually and physically withdrawn from the board and modified or configured to provide protection for the specific load. This can normally be done while the rest of the board is still energised.

Motor control centres

Motor controls are an integral part of the motor system design. There will be numerous large and small electrical motors offshore, each with different characteristics .Some motors have 'soft starts' to avoid an inrush of current at start-up that might damage motors and controls, while others have equipment designed to control speed.

Electrical control and protection

Key requirements of an electrical control system are that it will allow you to switch on the equipment and provide protection and disconnection during a fault condition.

The main types of protection are:

- Overcurrent: Overcurrent protection operates on a current overload and is usually almost instantaneous.
- Overload: Overload protection operates when an ongoing overload would cause overheating and damage to equipment. This is a time-dependent protection based on the severity of the overload.
- Differential: Differential protection operates when the phase current differential in a piece of equipment exceeds predetermined limits. It is intended to protect the equipment.

Protection devices

You should be aware of the following types of protection device which may be used in your projects, and I recommend an internet search if you require any more detail on principles or construction:

- Relay control.
- Air circuit breaker.
- Vacuum circuit breaker.
- Moulded case circuit breaker.

- Miniature circuit breaker.
- Fuse.

Voltage transformers

If you take a look at the SLD, you will see transformers.

For the purposes of this toolkit, think of transformers as "black boxes" which normally reduce higher voltages to lower voltages. Transformer design considerations for the electrical team include:

- Power ratings.
- Voltage ratings.
- Inrush current.
- Efficiencies.
- Physical size and weight.

Cables

Power cables distribute power around the facility. These are specified by the electrical discipline and can be either armoured or un-armoured. All external and some internal cabling is armoured to provide damage protection from dropped loads and other physical contact. Adding armour increases the difficulty and time to install and terminate the cables. Apart from the power and voltage ratings, they are usually specified to be either:

- Flame resistant: This cable is capable of continuing to transmit power under certain fire conditions.
- Flame retardant: Resists burning and typically has low smoke and emissions.

Installation of large high-capacity cables on an existing facility is a labour intensive task which can involve circa 15+ members in a team spaced out along a cable who are physically pulling it into position.

Trace heating

Trace heating is a special type of cable that provides heat and is used to prevent piping systems from freezing due to low ambient temperatures. This thermostatically-controlled trace heating cable is wrapped along piping systems that are at-risk and then covered with thermal insulation for heat conservation.

Lighting

There are three main types of lighting used in offshore installations, namely normal, emergency and escape lights.

Normal lighting: This covers all lighting that is supplied by the normal switchboards and provides general illumination whilst working or moving around the installation, internally or externally.

Emergency lighting: If power to the normal lighting system fails, the emergency power switchboard provides power to lights. In addition, many of the lights have battery back-up supplies that energise should the normal supply fail.

CHAPTER 24 ELECTRICAL

Escape Lighting: Escape lighting, as the name suggests, provides illumination along predetermined escape routes to the temporary safe refuge, helideck, and lifeboats. This often includes lighting at low levels, which will be more visible in a smoke-filled environment.

Low voltage distribution boards

Low voltage distribution boards supply a range of consumers around the facility including:

- Heaters.
- Power sockets, including welding sockets.
- Normal/emergency and escape lighting.
- Floodlights.
- Navigation aids.

Hazardous Area and Ingress Protection Ratings

Please refer to instruments (Chapter 23 Figure 23.8) for the hazardous area and IP ratings which are also used for instrumentation equipment.

Temperature class

Note that the maximum operating surface temperature of a piece of electrical equipment must always be below the ignition temperature of any potentially present explosive mixture.

Class	Hazards which will not ignite at temperature below:
T1	450°C
T2	300°C
T3	200°C
T4	135°C
T5	100°C
T6	85°C

Table 24.7 Temperature class table.

From the table shown above, you will see there is quite a bit of work involved in correctly and safely specifying electrical equipment.

Key Electrical Calculations

In addition to the overall electrical system, analysis calculations should be carried out for cables and lighting.

Cable calculations

Cables are normally referred to by the number of cores and the cross-sectional area of the conductors in mm^2. For example, 3C x 25mm^2 means a 3-core cable with each conductor having a cross-sectional area of 25mm^2. Cable size is selected as follows:

- Cable cross-sectional area is in mm^2 and is calculated based on current capacity.
- Allowable voltage drop is calculated and cable size is increased if necessary.
- Fault level withstand is calculated and cable size is increased if necessary.

Short circuit fault rating calculations

Under fault conditions, there are two major effects which the system must be able to withstand: The electro-magnetic effects which can translate into damaging mechanical forces, and the thermal effects due to a high current causing a rise in busbar temperature. Fault load calculations identify the maximum loads that can be permitted under fault conditions.

Lighting location and sizing/LUX levels

Different areas of the facility require certain minimum and maximum light levels which are established by standard calculations. These light/lux levels then translate into specifications and locations of lighting units.

Key Electrical Drawings

The electrical team will produce the following drawings to allow for specification of equipment, and for construction and commissioning to refer to. However, as mentioned earlier, the key document for providing an overview is the single-line diagram.

- Single line diagrams.
- Block diagrams.
- Wiring diagrams.
- Layouts.
- Schematics.
- Interconnects.

Electrical Equipment Package Management

Electrical engineering also manages specification generation and package management of electrical equipment, such as generators, transformers, switchboards, and motors, ensuring specification and design code compliance through manufacture, FAT, and SAT.

CHAPTER 24 ELECTRICAL

Electrical Safety Construction and Operating Considerations

Electricity, particularly HV systems, can be extremely dangerous, especially to untrained and unauthorised personnel. Offshore construction and connection of HV systems are carried out by fully-certified specialists and isolation or reinstatement of electrical systems is normally under the care of the platform Responsible Person Electrical (RPE).

Reproduced by permission of Quantum Controls Ltd

Fig 24.8 Electrical damage to a switchboard.

The above shows a switchboard that has been totally destroyed by a flashover.

Electrical Summary

The electrical discipline normally has a supporting role to the rest of the project unless it is a specific electrical system project or upgrade .I am always aware that the electrical team is dealing with high voltage at MW power levels to supply a small town and that safe design, construction, and commissioning is essential.

PETER F CRANSTON

CHAPTER 25

Technical Safety

Technical Safety is an engineering discipline which assures that engineered systems provide acceptable levels of safety. It focuses on safety in design, identifying and eliminating hazards in the design phase. This is often significantly more economical and practical than making changes later on when the hazards become real risks to people or assets. Technical safety deals with probabilities of catastrophic events, likely fatalities, and the design requirements to minimise these.

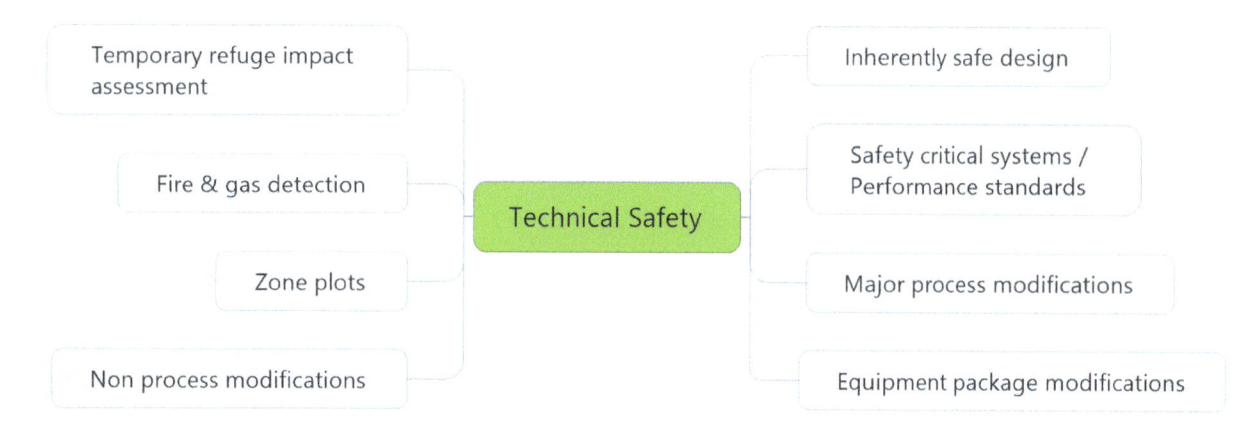

Figure 25.1 Technical Safety Mind Map.

This chapter will take you through the following main areas covered by technical safety:

- Inherently Safe Design.
- Safety Critical Systems and Performance Standards.
- Major Process Modifications.
- Equipment Package Evaluations.
- Non-Process Modifications.
- Zone Plots.
- Fire & Gas Detection.
- Temporary Refuge Impact Assessment.

Inherently Safe Design

As perfect safety cannot be achieved, common practice is to talk about inherently safer design, which is a design that avoids hazards instead of controlling them, usually by reducing the amount of hazardous materials and operations in the plant. An example would be to place process facilities on separate platforms away from the main accommodation. One of the key goals of the technical safety discipline is to ensure inherently safe design from the initial conceptual stage of a project.

Safety Critical Systems and Performance Standards

A safety-critical system or life-critical system is a system whose failure or malfunction may result in one (or more) of the following outcomes:

- Death or serious injury to people.
- Loss or severe damage to equipment/property.
- Environmental harm.

The components which make up safety-critical systems are known as safety-critical elements (SCEs) that must be designed to specified standards, reducing risks to acceptable levels.

A performance standard normally contains the following information about a safety-critical element:

- Functional requirements.
- Availability and reliability requirements.
- Required utilities.
- Roles and interfaces.
- Survivability requirement.

An example of a safety critical system would be the emergency shutdown system. Similarly, an example of a safety critical element within that system would be an emergency shutdown valve with a requirement to fully close within "X" seconds and surviving a jet fire for "Y" minutes.

Major Process Modifications

Typical examples of major process modifications would include a new subsea field connecting to an existing topsides facility, or reconfiguring gas compression equipment for declining reservoir pressure. The proposed design must be carefully evaluated to ensure that all hazards are identified and that risks are reduced to acceptable levels.

For this section, we will focus on FEED as this is where the most safety value-adding activities are carried out, and where changes can be made most cost-effectively.

PETER F CRANSTON

Safety screening report

Technical safety will use multi-discipline conceptual studies and previous specific technical safety studies to generate a safety screening report, which identifies the required studies in the FEED stage.

HAZOP

A hazard and operability study (HAZOP) is a structured and systematic examination of a complex planned or existing process or operation in order to identify and evaluate problems that may represent risks to personnel or equipment. A HAZOP is a group activity with a dedicated experienced HAZOP chair who takes a multidisciplinary team through the P&IDs in a structured manner, identifying hazards and action owners. The P&IDs will be split into smaller sections called "Nodes," and reference will be made to deviation guide words to assist in identification of risks. HAZOPs may last a number of weeks for large projects.

Parameter/Guidewords	Deviations
Flow	No, Less, More, Reverse, Misdirected
Pressure	Higher, Lower, No, Reverse, Vacuum
Level	Higher, Lower, Varying, Fluctuating
Temperature	Higher, Lower

Table 25.2 Sample HAZOP guide words.

HAZOPs will be carried out during studies, FEED, and detail design in increasing detail as P&IDs mature.

Cause and effects review

Cause and effects are covered in more detail in the instruments chapter, however, for now think of them as tables of instrumentation and control inputs that result in a set of outputs to control pumps, valves, alarms, etc. "Cause and effect" charts will be reviewed by the process team to ensure that alarms and executive actions do not contribute to unsafe conditions or incidents.

Safety Instrumented Function and Safety Integrity Level

A Safety Instrumented Function (SIF) is an instrumented function designed to maintain or achieve a safe state for a process. Its purpose is to:

- Automatically take a process to a safe state when specified conditions are violated.
- Permit a process to move forward in a safe manner when allowed during specified conditions.
- Take action to mitigate the consequences of a process hazard.

A Safety Instrumented System (SIS) is a combination of equipment or logic solvers designed to implement a SIF.

Safety Integrity Level (SIL)

Safety integrity level (SIL) is defined as a relative level of risk-reduction provided by a safety instrumented function or to specify a target level of risk reduction. In simple terms, SIL is a measurement of performance required for a safety instrumented function (SIF). Four levels of SIL are defined, with SIL 4 having the highest level of integrity and SIL 1 the least. Typically, in an offshore environment the highest integrity requirement will be SIL3. The design should aim to minimise high SIL requirements through the use of inherently safe design as higher SIL ratings mean higher project costs.

Fire & Explosion Analysis

This evaluates the impacts of various scenarios of fire and explosion on people, platform structure, and facilities and is considered at the conceptual study phase. The analysis will determine fire and blast design accidental loads for the FEED and detail design phases.

Jet and pool fires

A liquid/gaseous hydrocarbon release can result in a pool fire or the much more aggressive jet fire. A hydrocarbon pool fire will generate temperatures up to 1300 Celsius within ten minutes of ignition, with heat fluxes of around 200 kW/m^2. A jet fire will exhibit the same temperature rise, but the heat flux could double that of the pool fire [Ref11]. A jet fire impinging on unprotected structural steelwork can cause structural failure within minutes. Jet fires can be modelled to determine duration, extent, and consequences, allowing suitable active or passive protection to be specified.

Reproduced by permission of Gexcon UK

Figure 25.3 Jet fire at test facility.

Jet fire simulation

Sophisticated computational fluid dynamics (CFD) software can now model jet fires based on leak location, pressure, and direction (as shown in the ruptured vessel of Figure 25.5). This allows for exploring various leak and ignition scenarios, calculating heat fluxes, and developing mitigations.

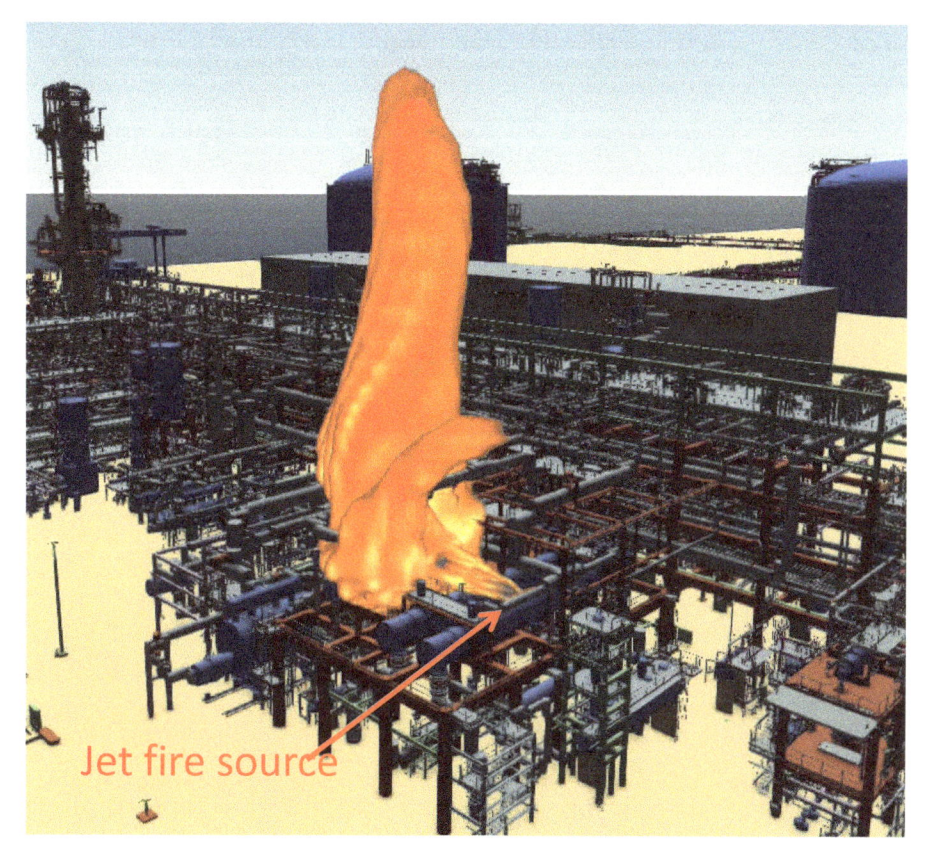

Reproduced by permission of Gexcon UK

Figure 25.4 Jet fire simulation using FLACS™.

Gas Dispersion Analysis

This analysis is carried out using industry standard software packages such as CFD to evaluate the potential size of a gas cloud from various leak sources or vents under different wind strengths and directions.

Active and passive fire protection

Active fire protection covers active systems such as automatic water deluge on "at-risk" process plants, automatic water mist fire suppression under turbine hoods, and numerous fixed and mobile water and foam fire-fighting equipment.

Passive fire protection (PFP) refers to coating and shielding systems that physically protect equipment or accommodation modules from temperature increases for a set period of time under specified fire conditions. PFP can be built into rigid panels, take the form of quilted jackets or can even be a spray or trowel-applied "plaster" of a specified thickness.

The purpose of PFP is to:

- Prevent load-bearing structural steelwork from reaching 400°C, the temperature at which steel strength is reduced by half.
- Protect process vessels.
- Prevent heat transfer through walls into habitable spaces by limiting the inner wall temperature.

There are three fire categories against which PFP coatings are specified:

- 'A' rating: Cellulose fire.
- 'H' rating: Hydrocarbon fire.
- 'J' rating: Jet fire

There is also a time component associated with the coatings fire rating, i.e. a J30 rating provides protection in a jet fire for 30 minutes.

Blast overpressure studies

Using gas cloud sizes calculated using CFD, blast explosion overpressures can be determined. Explosion blast overpressures are considered significant if greater than 0.2 bar (2.9 psi). Whilst this seems low, the forces can be very large when applied over a large area. Critical structures and equipment/piping supports may need to be designed to withstand blast overpressures and drag loads to prevent collapse and escalation of an incident.

Reproduced by permission of Gexcon UK

Figure 25.5 Blast overpressure plots.

The different colours in Figure 25.5 correspond to generated blast overpressures and can be used to evaluate loads on structures, piping, and equipment in order to design suitable restraints for the blast scenario.

Flare radiation modelling

Analysis is carried out on the radiation from the platform flare under different blow down and wind conditions. Results from these studies may show that additional personnel and equipment thermal protection is required.

Reproduced by permission of Gexcon UK

Fig 25.6 Flare radiation simulation

The above simulation shows the temperature (K) of the flare and the radiated heat energy per square meter (kW/m^2).

Quantitative risk assessment

The quantitative risk assessment (QRA) is a speciality field within technical safety that is often subcontracted and is primarily concerned with determining the Individual Risk Per Annum (IRPA) and potential loss of life (PLL) caused by undesired events. It includes both initiating events and various escalation scenarios. Specialist software is often used to model the effects of such an event and calculate the IRPA and PLL in order to demonstrate that risks are minimised within accepted industry norms.

Escape evacuation and rescue analysis

Escape Evacuation and Rescue (EER) evaluates risks to escape routes, and escape/evacuation critical systems. Escape/evacuation critical systems include TEMPSC (Totally Enclosed Motor Propelled Survival Craft), life-rafts and escape to sea devices, such as knotted ropes and Donut™ systems. The fundamental requirement is to ensure that escape/evacuation routes will remain effective during a major accident hazard (MAH) incident and to demonstrate adequate routes and timings for movement of personnel to escape locations.

Life-saving equipment

Technical Safety is responsible for specifying the following life-saving equipment:

- Portable fire extinguishers.
- Evacuation grab bags including smoke hoods.
- Fire hoses.
- Lifejackets.
- Escape signage.
- Escape ladders.
- TEMPSC.
- Liferafts.

Other specific responsibilities

Other Technical Safety scope responsibilities include:

HAZIDs (Hazard Identification) is a qualitative technique for the early identification of potential hazards and threats affecting people, the environment, assets, or reputations.

ENVIDs reviews enable the team to identify environmental aspects that come about due to an interaction between the facility and its surroundings in order to plan for, avoid, or mitigate potential adverse impacts, such as unplanned releases, process wastes, and construction wastes.

MAH Register is a report detailing identified MAH and safeguards/barriers. In UK waters, the regulations governing MAH are contained in The Offshore Safety Directive (OSCR 2015). The primary aim of OSCR is to reduce the risks from major accident hazards, (MAH), to the health and safety of the workforce employed on offshore installations or in connected activities. The regulations also aim to increase the protection of the marine environment and coastal economies against pollution and ensure improved response mechanisms in the event of such an incident.

Safety Case Impact Assessments are conducted by Technical Safety to review proposed modifications for an existing facility. The purpose is to determine whether any of the safety and environmental-critical elements are affected and a material change is required under OSCR 2015.

ALARP Report stands for "As Low as Reasonably Practicable," and is the term often used in the regulation and management of safety-critical and safety-involved systems. The ALARP principle states that residual risk shall be reduced as far as reasonably practicable i.e. within reasonable cost.

Equipment Package Evaluations

Technical Safety supports the equipment package engineers in ensuring the following equipment evaluations have been carried out:

- Identification of potential ignition sources from heat or sparks.

- Identification of equipment shutdown requirements.
- Noise and vibration assessments on equipment packages.
- PUWER assessments ensuring that operator access for use and maintenance is both safe and ergonomically acceptable.

Provision and Use of Work Equipment Regulations

The Provision and Use of Work Equipment Regulations (PUWER) applies to all equipment used by an operator to ensure that the equipment is suitable for its purpose, maintained to be safe, not risk health and safety, and has been inspected by a competent worker who has recorded the results. Simply put, it is a tool to ensure that the equipment provided by a supplier is safe.

In practice, formal PUWER reviews will be carried out using a standard checklist after the general arrangement drawings have been completed for a piece of equipment i.e. when major items have been installed, and the equipment has been installed offshore and is ready for use. This staged method means that significant design change requirements will be picked up early in design when costs to change are relatively low.

Non-Process Modifications

Areas covered under this section include:

- Construction HAZID.
- Dropped object studies.
- Human factors.
- Performance Standards.

Construction HAZID

This qualitative technique looks at specific hazards that may occur in construction and allows the design to be modified at an early stage to accommodate these hazards. An example of a construction hazard is welding while the platform is live, with the solution being to do the hotwork during a shutdown or inside a pressurised habitat.

Dropped object studies

Whilst the probability of a dropped object may be low, the consequence of a dropped object on topsides equipment/piping or onto subsea infrastructure could result in a major incident. Potential dropped objects will be evaluated to determine the potential energy of dropping and damage that would result. Mitigations include physical protection, bumpers and scaffolding, and shutdown/hydrocarbon-freeing of equipment and pipework during lifting operations.

Human factors

Human factors in topsides engineering is a speciality sub-division of technical safety which has recently obtained more focus. Designing your platform using onerous design codes and the most reliable equipment only reduces risk so far. We must also account for the most unreliable part of the system: Humans. In an incident, the associated physiological-stressed humans may not always react as planned.

Human factors, also known as ergonomics, is essential in ensuring that designing for human operation incorporates reducing risk and probability of error and suboptimal operation. Human factors is primarily concerned with the study of how humans react physically and psychologically to a particular scenario or incident, allowing layout systems and equipment to be designed for the safest, most effective, and lowest stress use by humans. A typical example would be to layout a facility control room with its associated alarms and graphics.

Good design would typically provide as much usable information as possible without swamping the operator with much, if any, unusable data.

Performance standards

Performance standards are required to be developed for identified safety and environmental critical elements. A performance standard is a statement expressed in qualitative or quantitative terms of the performance required of a system or item of equipment, and is used as the basis for managing the hazard (e.g. planning, measuring, controlling, or auditing) throughout the life cycle of the installation.

Note: The regulations do not specify what the performance standards should be; the duty-holder must decide, taking into account the circumstances on the particular installation.

Platform Hazardous Area Zone Plots

The categorisation of hazardous areas in a facility is carried out by Technical Safety and is driven by the location of flanged piping joints, vessels, and vents. Inputs to this process include piping routing isometrics, general arrangements, and the hazardous area classification schedule. The output of this process is a hatched area drawing showing the extent of each zone, which are defined as follows:

- Zone 0: Explosive atmosphere is present continuously or for long periods.
- Zone 1: Explosive atmosphere is likely to occasionally occur during normal operation.
- Zone 2: Explosive atmosphere is not likely to occur during normal operation, or only for a short period.
- Non-hazardous: Explosive atmosphere is not expected to be present.

An example of a reason for a zone 2 area would be the presence of a hydrocarbon line with flanges transiting through an otherwise non-hazardous area. If the same line was fully welded with no joints, it could then be classed as a non-hazardous area.

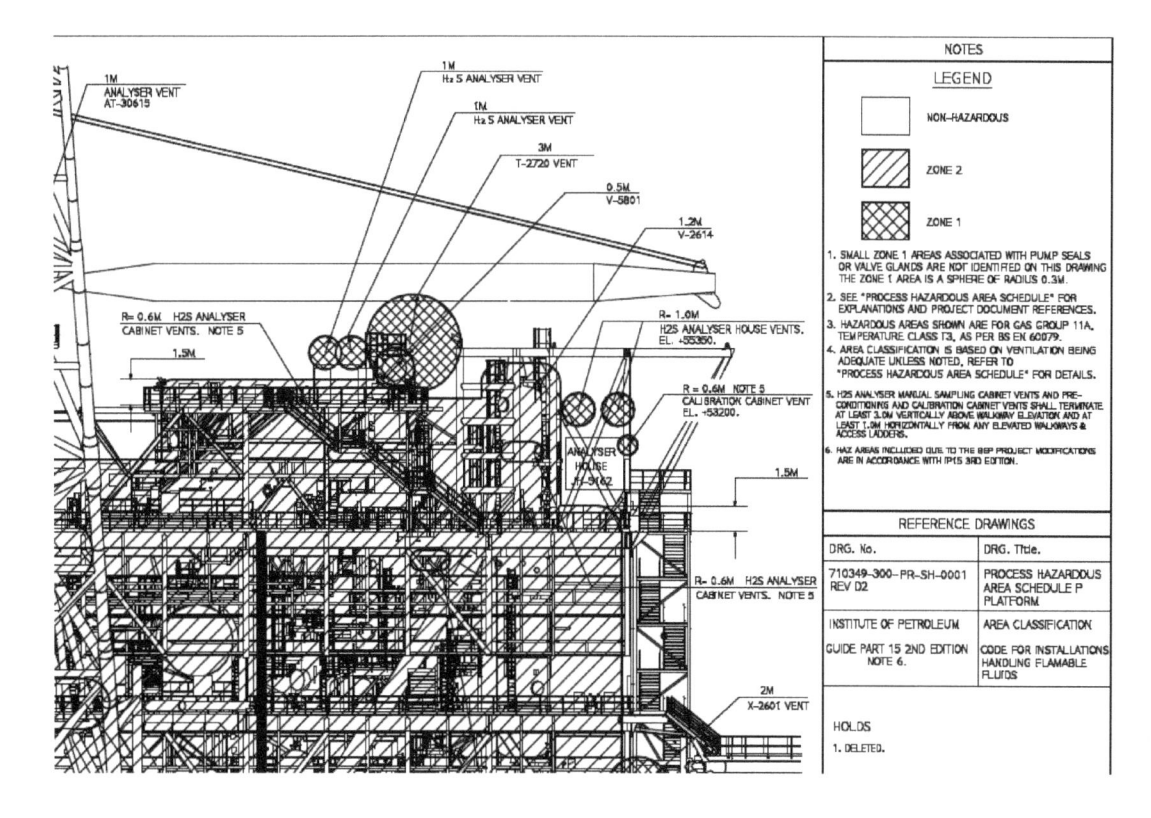

Figure 25.7 Hazardous area drawing excerpt.

It can be seen from the type of hatching on the excerpt that the majority of the elevations shown is zone 2, with zone 1 spheres located around vents. It can also be seen that the access stair at the right-hand side is a non-hazardous area.

These hazardous area drawings will also be used by electrical and instruments to ensure that equipment for a particular area is correctly rated with respect to Zone requirements.

Fire & Gas Detection

Technical safety is typically responsible for the overall fire and gas philosophy for the process plant and the accommodation/temporary refuge. Until recently, detection was based on applicable legislation and client and industry standards and specifications. Currently, detection is expected to reflect the findings of gas dispersion analysis and setting performance standards, translating into specifications and detection sensor layouts which is sent to instruments for integration into the overall fire and gas system.

Temporary Refuge Impact Assessment Offshore

As offshore oil and gas installations are remote and often in inhospitable areas of sea, special consideration is given to providing a Temporary Refuge. This is usually in the main accommodation location of the platform that is designed to withstand all reasonably foreseeable incidents and will protect personnel from fire, blast, smoke, and gas.

Aligned with the platform emergency shutdown system, the accommodation is designed to provide life support for a specified period of time until evacuation by helicopter or sea is carried out.

As a result of the Piper Alpha tragedy in 1998 in which 167 people lost their lives, legislation was implemented specifying minimum requirements for platform Temporary Refuges.

Technical Safety Summary

Technical Safety covers almost every aspect of platform design to ensure that facilities have reduced the risk of an incident to an acceptable frequency, and should an incident occur, it does not escalate with personnel being able to escape from the incident to the temporary refuge or evacuate from the platform.

Routes to Becoming a Technical Safety Engineer

Most Technical Safety Engineers have an MSc in Safety Engineering, Reliability, and Risk Management. However, one may have many years of industry experience instead.

"No job is too important that we cannot take the time to do it safely".- Unknown

PETER F CRANSTON

CHAPTER 26

HSSE

HSSE stands for Health, Safety, Security and Environment, but first let's ask, why is HSSE so important? It is often the first agenda item at many meetings, briefings, and presentations. There are a number of answers, including:

- We do not wish any aspect of our projects to damage peoples' short- or long-term health.
- Safety is paramount and we do not wish to injure people or cause damage to equipment and facilities.
- We do not wish to expose personnel to security risks.
- We do not wish to damage the environment.

HSSE covers all aspects of a project, from concept to offshore construction and commissioning. It includes the entire workforce, such as the client, main contractor, subcontractors, and suppliers.

Finally, there is a commercial aspect to HSSE. When tendering, contractors are evaluated on their HSSE performance and if it is significantly lacking, they will fail the pre-qualification and will not be allowed to tender.

Figure 26.1 HSSE Mind Map.

Health and Hygiene

Workforce fitness

A healthy workforce is a productive workforce. Occupational health, for which there is often a dedicated nurse or department, is concerned with ensuring that:

- Personnel are physically and mentally fit to carry out their roles.
- Exposure to operational hazards are minimised or controlled.

For personnel working onshore and offshore, stringent medical examinations are carried out to ensure fitness. Medical evacuation from an offshore facility is not always immediate due to weather limitations and a relatively low-risk injury onshore can quickly become life-threatening offshore. Within the oil and gas industry, drug and alcohol testing is carried out routinely as part of medical examinations or if there is a just cause. Alcohol and substance abuse and hazardous environments do not mix.

Occupational health risks

Depending on a person's job and the equipment and substances that he/she will be exposed to, personnel are assessed and monitored on:

Noise level exposure	Manual handling
Vibration exposure	Working position (i.e. wrist, back, neck)
Hazardous substances (i.e. asbestos)	Repetitive strain
Exposure to radioactive isotopes	Hot/Cold, Ultraviolet etc. exposure
Mental health.	

Table 26.2 Occupational health risks.

Assessing and monitoring will ensure that exposure to risks are minimised.

Legislative requirements

There is specific legislation relating to all areas or health risk exposures, including:

- Noise.
- Vibration.
- Radiation.
- Display Screen Equipment.
- Lifting.

The principles for assessment and risk minimisation for each are similar. An example is shown below for CoSHH.

Control of hazardous substances

In the UK, CoSHH (Control of substances hazardous to health) assessments focus on the hazards and risks from hazardous substances in the workplace and determines how to eliminate or reduce them. There are similar legislations in other jurisdictions as well that may go under different names. There are 5 steps in managing hazardous substances:

- Identify hazards i.e. anything that may cause harm.
- Decide who may be harmed, and how.
- Assess the risks and take action.
- Make a record of the findings.
- Review the risk assessment.

All hazardous materials will need to be assessed before they are specified, shipped, and used onsite. Offshore Assets build databases of all the hazardous substances that can be used on-site, adding new items as required.

Safety Leadership and Accountability

The safety culture in a project is set by the Project Manager, senior management within the organisation, and middle management offshore that is directly supervising the construction workforce. The aim is to support a strong safety culture in the project and with the subcontractors. Project Managers control the project resources, so they are the driving force who are supported by HSSE advisors. It should be noted that HSSE advisors are not accountable for safety, but rather that both Project Managers and senior organisation managers can be held legally accountable, facing criminal charges for unsafe activities that result in injury.

Risk Assessment Management

Employers have a legal duty to protect the health, safety and welfare of their employees and other people who might be affected by their business, meaning that making sure workers and other personnel are protected from anything that might cause harm. To minimise the risk to personnel, the public, and the environment, risk assessments should address all hazards in the workplace.

Hierarchy of control

Risks should be reduced to the lowest reasonably practicable level by taking preventative measures in order of priority. The table below sets out an ideal order to follow when planning to reduce risk from design and construction activities. When reviewing or undertaking a risk assessment, close reference should be made to the hierarchy of control to satisfy the safest control measures. Please see the following figure:

Hierarchy of Controls

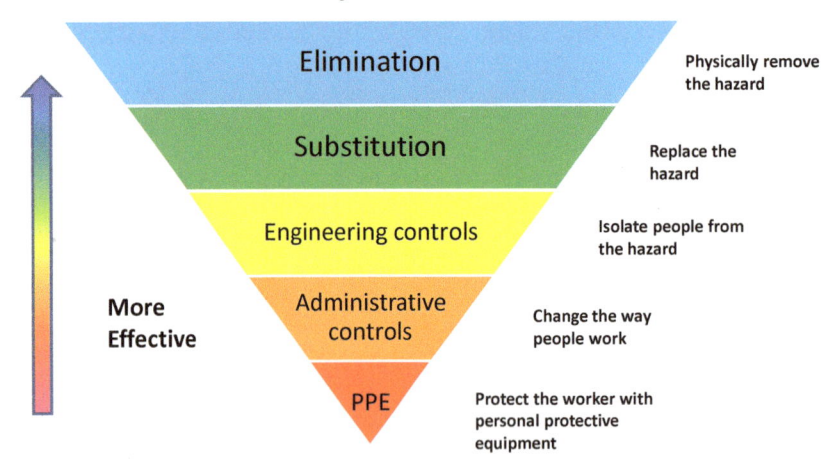

- **Elimination** — Physically remove the hazard
- **Substitution** — Replace the hazard
- **Engineering controls** — Isolate people from the hazard
- **Administrative controls** — Change the way people work
- **PPE** — Protect the worker with personal protective equipment

More Effective ↑

Figure 26.3 Hierarchy of controls.

Sub-Contractor Management

Sub-Contractors are an important part of extended project teams and it is important to make sure that we understand the potential risk that subcontractors can pose and assess them accordingly prior to engaging on-site. Contractor pre-qualifications are typically ranked on their submissions according to the following categories:

- 🟢 Meets all requirements and is an approved, pre-qualified contractor and therefore can proceed into pre-mobilisation and mobilisation phases.

- 🟡 Develop a Support and Engagement Plan that adequately addresses the required areas for improvement, verified and stewarded by the Project Management Team.

- 🟠 Contractor not to be selected for bid list or execution. Only exceptions are where services are single-sourced with no alternatives or other contractual or geographical stipulations (i.e. "local content").

Safety Triangle

The safety triangle is a simple yet powerful aid which illustrates the relationship between the quantity of unsafe behaviours and serious injuries or fatalities

This tool was developed by Herbert Heinrich in the 1930s[Ref12] and refined over the years, showing that for every accident resulting in a fatality or major disabling injury, there are approximately 300 unsafe incidents associated with it. The key is to identify and reduce the number of near-misses and at-risk behaviours to reduce the probability of a fatality.

Safety Triangle

- 1 Fatality
- 30 lost day cases
- 300 recordable injury cases
- 3,000 near misses
- 300,000 at-risk behaviours

Figure 26.4 Safety triangle.

Proactive Risk Reduction

Playing our part in risk reduction

We can all play our part in day-to-day HSSE both in the office and onsite by using the following examples of simple proactive activities:

Safety moment

A safety moment is a brief safety talk about a specific subject at the beginning of a meeting or shift and especially when reaching significant points in a project, or when a number of new subcontractors are brought together. It establishes that the seriousness of safety in the project. A safety moment will typically present a project-relevant good or bad practice, such as an accident review, audit findings, trend review or issue, and generate discussion around the topic. These safety moments need only take five to ten minutes.

Including a Safety Moment at the beginning of your meeting can help bring safety issues or topics up in a timely, clear, brief, and non-threatening way, reinforcing safety knowledge and everyone's commitment towards a positive safety culture.

Tip: Mentally record any attendees who are not fully engaging or are negative about the "safety moment" as they likely need more support to ensure that their department or organisation delivers a good safety performance. In addition, discuss potential safety audits with your HSSE advisor.

Safety conversation

A safety conversation can be carried out at any time and is most effective when both observer and recipient have had the appropriate training. The observer will approach the person working and start a safety conversation in a non-adversarial manner about what has been observed. If a negative behaviour has been witnessed, the observer will seek to get acknowledgement from the recipient of the risks alongside a verbal commitment to rectify the behaviour. A safety conversation can be also used to praise and reinforce positive safety behaviours.

Stop card/hazard observation card

All employees have the obligation to stop work anytime they feel that their safety or the safety of other employees is at risk. In a mature organisation, there should not be any push back if you intervene politely and advise that you have noticed an unsafe action or condition.

Organisations may have a card, and often also an IT-based system, in which any employee or subcontractor can raise and submit a card that identifies a hazard, an unsafe behaviour, or an environmental issue. Data from these cards can be analysed and generic risk issues and remedial actions can be identified as a result. Data from these cards can also help form the base of the safety triangle. Personnel should be encouraged to raise cards to report good safety behaviours rather than just the unsafe issues.

Safety Alerts

In many operating locations, offshore facility operators and government agencies share safety incident alerts to minimise the risk of a particular event happening elsewhere. They do not name the organisation or the asset, allowing for open sharing of the actual incident details. These are normally cascaded by the project HSSE advisor within the project and, if considered a serious enough alert, he/she will call an immediate meeting to brief the team with the support of project management.

Note: Safety alerts and safety moments can be found on the Step Change website, https://www.stepchangeinsafety.net

---

HSSE Reports

HSSE reports are usually compiled and presented to the project management team and client management on a weekly, monthly, or quarterly basis. The reports are highly visual and project-specific documents showing target and actual numbers in the following categories:

- Serious injury/fatality.
- Lost time incident.
- Medical workcase.
- Near misses.
- Unsafe behaviours.
- HSSE actions that can impact the project.
- Actions close-out status.

Leading and lagging indicators

Leading and lagging indicators are used on a project to both prevent reoccurrence of injuries and to prevent potential injuries.

Leading indicators

A leading indicator is a measure preceding or indicating a future event used to drive and measure activities carried out to prevent and control injury. Examples include:

- HSSE training programme and targets.
- Number and quality of stop cards raised.
- Employee perception surveys.
- HSSE audits.

Leading indicators are focused on future safety performance and continuous improvement.

Lagging indicators

Lagging indicators measure a company's incidents in the form of past accident statistics. Examples include:

- Injury frequency and severity.
- Legislative recordable injuries.
- Lost workdays.
- Worker's compensation costs.

The major drawback to only using lagging indicators of safety performance is that they tell you how many people were injured and how bad the injuries were, but not how well your company is doing at preventing incidents and accidents from happening.

Security

Personal security hazards may be an issue in some locations, so HSSE will liaise with specialist security advisors to ensure that risks are minimised during travel or when onsite. Security measures may involve provision of escorts or security upon arrival at certain locations, or in the most serious security situations involve repatriation of personnel from the affected area.

Tip: When considering travel to an area that you are uncertain about security, check your home countries foreign office travel guide website, such as https://www.gov.uk/foreign-travel-advice for recommendations and advice. If the office does not recommend travel refrain from going.

Environmental

Licensing

Protection of the environment, air, water, and seabed is legislated in most jurisdictions, with permits being required for various operations and planned discharges within certain limits, such as flaring of gas and discharge of treated, oily water back into the sea, amongst others. Licenses are issued by government agencies and are held by oil and gas installation operators

Pollution prevention

The emphasis offshore is on pollution prevention, which can be achieved by design, such as with containment bunds around pumping systems, or by material selection, such as water-soluble and environmentally-friendly lubricants.

To give an example of how seriously pollution is taken, helicopter pilots transferring personnel on- and offshore in the North Sea are required to notify authorities if they spot any oil sheens on the surface of the sea. This will immediately trigger an alert and require the offending facility operator to take corrective action, if none had already been taken.

Waste management

All waste is carefully controlled in the offshore environment to ensure that it is correctly sorted and stored for future onshore treatment and disposal.

HSSE Advisor Responsibilities

HSSE advisors are responsible for:

- HSSE management systems development.
- Auditing and supporting subcontractor HSSE activities.
- Preparing and updating project emergency response plans and communicating with all emergency contacts.
- HSE input into design to ensure harm-free engineering.

Legal Implications

In the UK, the Health and Safety Executive (HSE) provides both guidance and legislation enforcement regarding HSSE breaches. If there is a breach, they will normally issue an improvement notice to the operator advising of the shortfall and requesting remedial action within a stated time frame. If remedial action is not achieved, they can issue a prohibition notice on the operating part or all of the installation. Oil and gas company management "sit up" when an improvement or prohibition is mentioned because it affects their production targets and balance sheets.

Behavioural Safety: A Significant Recent Change in Safety Management

There has been a large uptake of "behavioural safety" approaches over the past decade in the oil and gas industry. These approaches are based on the premise that a significant proportion of accidents are primarily caused by the behaviour of people rather than equipment failure. Simply put, it includes:

- Definition of safe and unsafe behaviours.
- Observations of behaviours by trained observers.
- Feedback and reinforcement of safe behaviours.

Whilst behavioural safety is part of safety management, it must not be perceived as an alternative to ensuring that adequate engineering design and effective safety management systems are also in place.

HSSE Summary.

We need to remember that all employees are someone's father, wife, son, daughter, etc. How would these people feel if their loved ones failed to come home due to a work-related accident? It is incumbent on everyone at all levels to look after their own and their colleague's safeties Apart from the human cost, good safety also makes good financial sense.

"Tomorrow - your reward for working safely today." - Robert Pelton

CHAPTER 27

HVAC

HVAC (Heating, Ventilation and Air Conditioning) is used in many areas on offshore installations. Often dismissed as "wind and tin," we will see in this chapter that HVAC systems perform vital functions in not only keeping process areas safe but also providing accommodation areas that are comfortable and habitable.

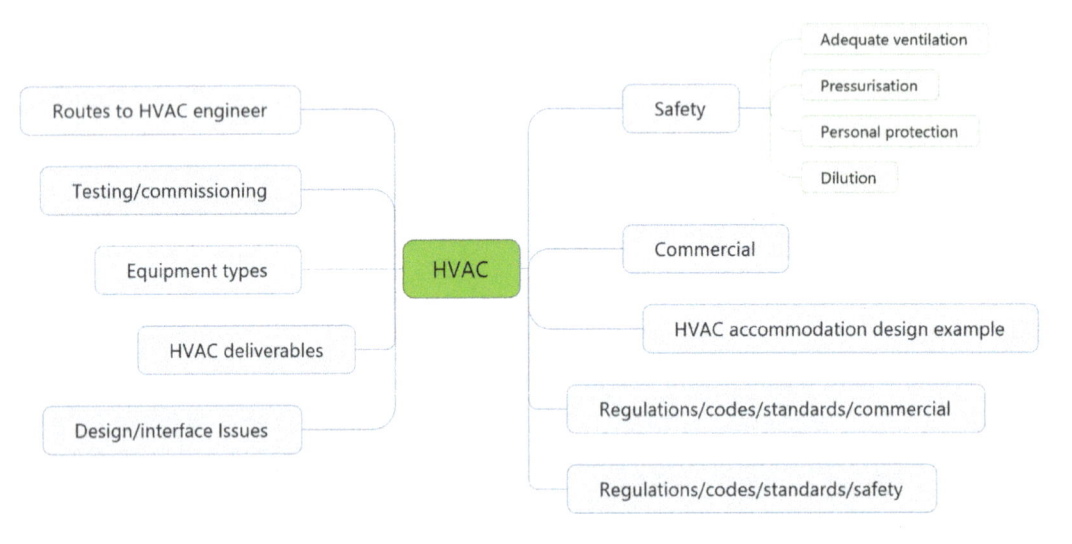

Fig 27.1 HVAC Mind Map.

HVAC Applications

The two main applications of HVAC are:

- Safety HVAC systems.
- Commercial HVAC Systems including offshore accommodations and offices.

Safety HVAC Systems

Safety HVAC systems are designed to protect equipment and personnel from a build-up of explosive or noxious gas and vapour mixtures and are subdivided below:

- Adequate ventilation systems.
- Pressurisation HVAC systems.
- Personnel protection HVAC systems.
- Dilution HVAC systems.

Adequate ventilation

Adequate ventilation is a term defined by hazardous area codes from the American Petroleum Institute and the UK Energy Institute. Typically, all process areas are required to be adequately ventilated to ensure that there are no build-ups of gas clouds from fugitive leaks. In congested modules, it is quite common to see HVAC nozzles directed at ventilating potential dead spots.

Fig 27.2 HVAC nozzle.

Later generation process equipment modules are often open, allowing for natural ventilation and reducing or eliminating the requirements and associated costs and weight of HVAC plant.

Fig 27.3 Open module with natural ventilation.

Pressurisation

In positively pressurised systems, air from a non-hazardous area is introduced to an enclosure containing spark-potential equipment or naked flames and is over-pressurised to prevent hydrocarbon gases from entering. Typically, electrical switch-rooms, local equipment rooms, workshops, laboratories, and accommodation modules require pressurisation.

Personnel protection

In this application, air is used to divert the hazardous vapour or dust away from personnel with air velocities of between 0.5 to 1.0 m/s, depending on the specific application.

Personnel-protection HVAC can be found on:

- Fume cabinets.
- Drilling mud tanks.
- Drilling shale shakers.
- Welding bays.

Dilution

Dilution is a special type of ventilation mainly used when there are hot surfaces and flammable gases in the same location, such as gas turbine enclosures where surface temperatures may reach 1350°C. These systems introduce so much air that any escaping gas cannot ignite because it is diluted. Typical air flows for turbine enclosures are 120 air changes per hour.

Commercial HVAC Systems

Commercial HVAC systems are more aligned to building service principles and are typically used for offshore accommodation and offices. These areas require HVAC for a number of purposes including:

- Providing clean air from a non-hazardous area of the platform.
- Providing a pressurised module to prevent ingress of flammable or toxic gases.
- Conditioning air to the correct temperature and humidity.
- Allowing for individual comfort control in sleeping cabins.
- Extracting "used" air from galleys, toilets, wet-rooms and smoking rooms.
- Emergency shutdown and closure of HVAC dampers in the event of smoke or gas detection.

Accommodation HVAC Design Example

The following example covers the design, procurement, and construction process of a new HVAC system for a new accommodation module which interfaces with an existing accommodation module in the facility.

Accommodation HVAC principles

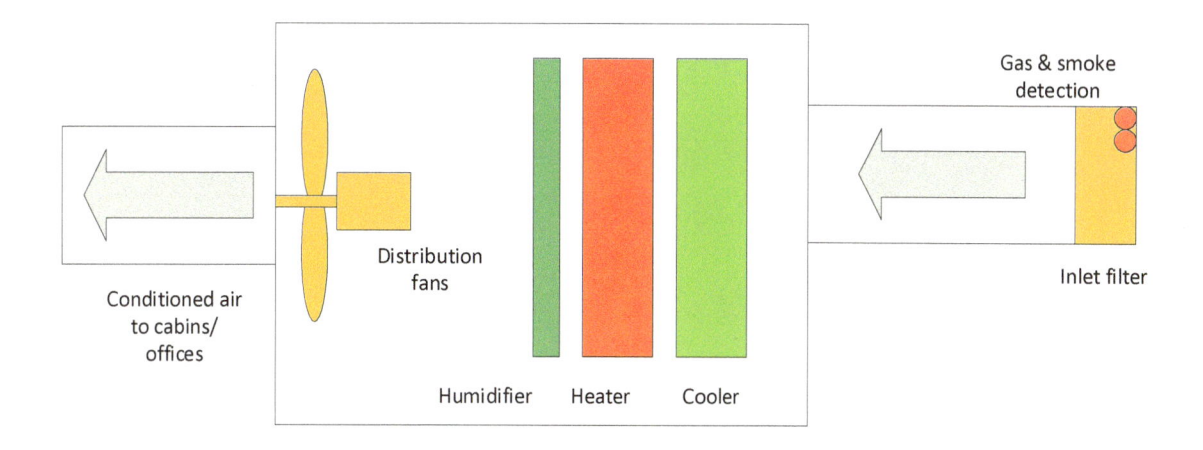

Fig 27.4 Simplified block diagram.

CHAPTER 27 HVAC

First, the basic principles. Clean air from a non-hazardous area is filtered and ducted to the supply HVAC skid where the air is cooled to knock out water, then heated to the required temperature and finally humidified by the injection of steam. This conditioned air is then distributed to cabins.

Accommodation conceptual study

During the conceptual study, various layouts and location options for the accommodation will be generated and evaluated. Once a preferred concept is agreed upon, the FEED will be carried out and HVAC will be tasked with completing the HVAC feed input.

HVAC FEED input

The starting point for HVAC is the accommodation layout, dimensions, and volumes of the internal areas.

The HVAC discipline will carry out calculations to determine the demand flow rate for the system, taking into account internal accommodation volumes, internal and external leakages, and required pressure differential. HVAC will also carry out an overall heat balance to ensure that sufficient heating and cooling capacity is available to cope with maximum and minimum external design temperatures of 0°C to 26°C. The following general specifications apply in the UK sector:

- Internal cabin temperature is thermostatically controlled and nominally set to 18°C.
- A minimum of 12 litres/second of clean, conditioned air per person.
- A minimum of 5 room air changes per hour, with more in bathrooms and smoking areas.
- Relative humidity set to a minimum of 40%.

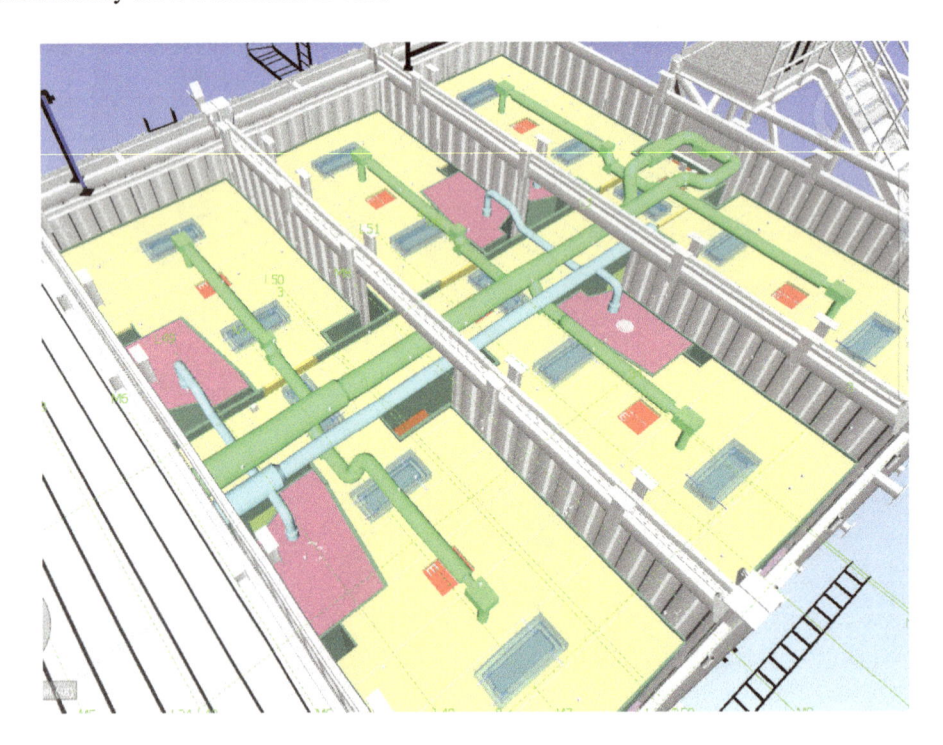

Fig 27.5 HVAC internal ductwork 3D layout.

PETER F CRANSTON

Internal ductwork and equipment

In parallel with the engineering calculations, the HVAC designer will locate the main items of the HVAC plant and will also route the ductwork to and from supplies and extracts. Modules are normally designed using a multi-user 3-D CAD (Computer Aided Design) system. In an accommodation module, design priority is normally given to routing HVAC ductwork as it is the largest item. Ductwork is normally located at high levels above cabin or office ceiling, with vertical duct shafts in between levels (See Figure 27.5).

Layout design of equipment and ductwork will be carried out in FEED to allow sizing of equipment and ductwork for HVAC costs to be accurately determined and included in the project estimate. Any major clashes of equipment between HVAC and other disciplines will be identified and resolved in FEED. Finally, depending on the overall project schedule requirements, some HVAC packages may be ordered during FEED.

External equipment

The main supply and extract skids are commonly located on the roof of the accommodation module (shown in the figures below) and connected to large cross-sectional area supply and extract ductwork. You can see in Figure 27.7 the seemingly long and convoluted inlet duct route and length. The reason for this is that the length is sized to allow gas and smoke detection and full closure of fire dampers before the contaminated air reaches the accommodation. This is typically a time of 8 to 10 seconds.

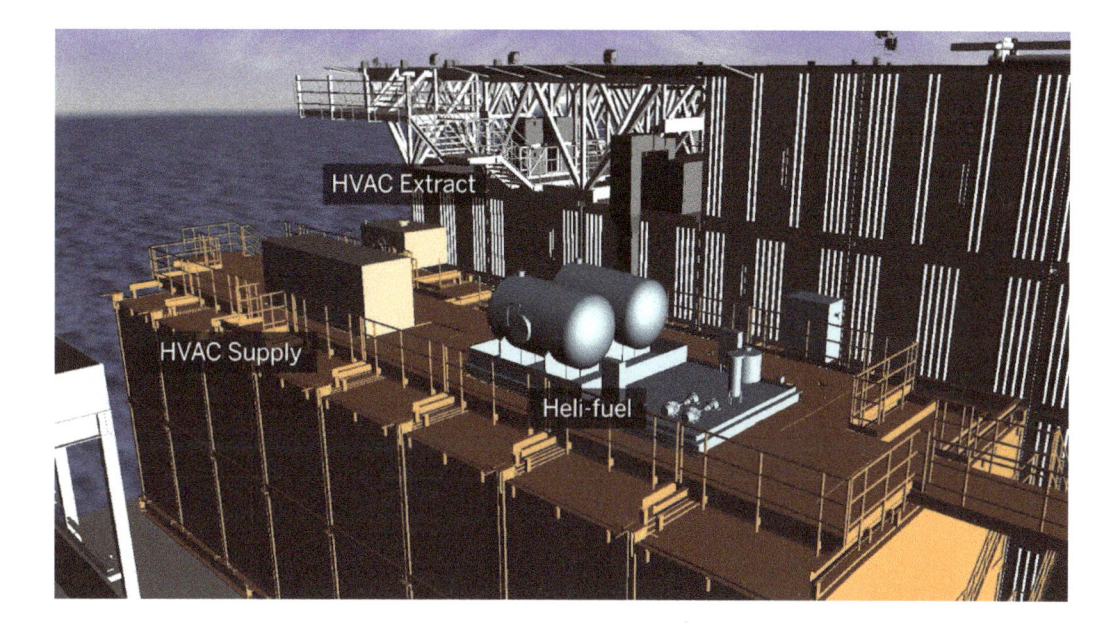

Fig 27.6 HVAC external equipment located on module roof

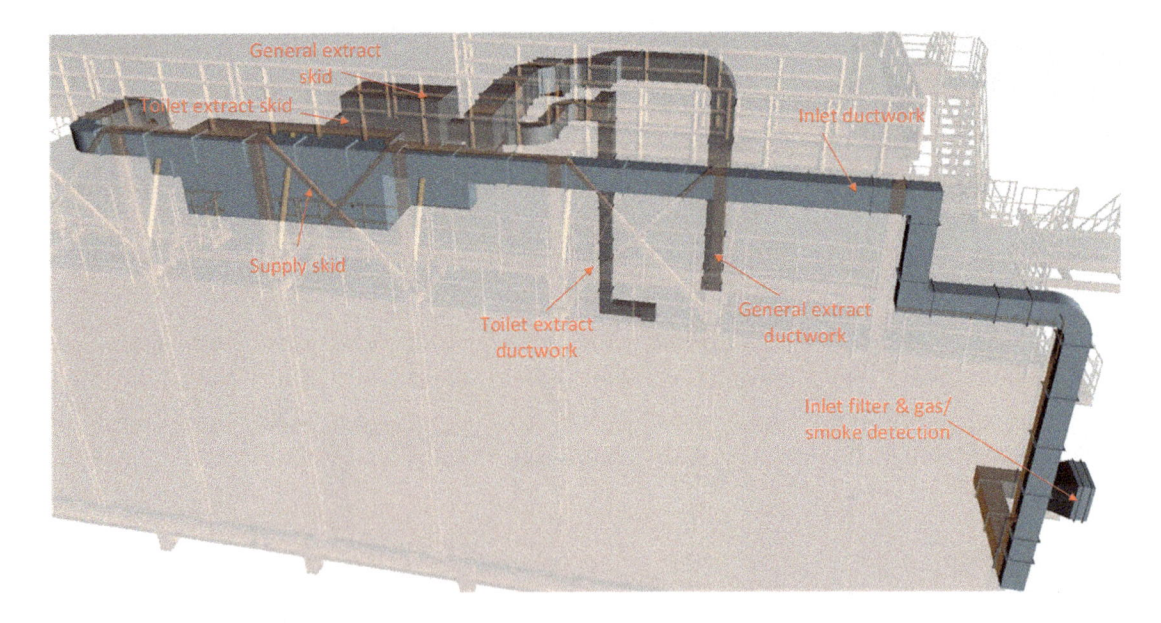

Fig 27.7 HVAC external ductwork routings.

Commercial Regulations, Codes, and Standards

Commercial HVAC system design and specifications are based on the following guides and standards:

- CIBSE Guide B.
- ASHRAE Handbook HVAC Systems and Equipment.
- ISO 15138 Offshore production installations, heating, ventilation and air-conditioning.
- Client-specific specifications.

Safety Regulations, Codes and Standards

The primary requirements for safety-based HVAC systems are derived from legislation in the form of statutory instruments, with the main ones being:

- Safety Case Regulations.
- PFEER Regulations.
- ATEX Regulations.
- CoSHH Regulations.

Design and Interface Issues

The following elements will be considered during the design and specification of an HVAC system:

- Designing for a saliferous (salt laden) environment primarily using stainless steel for ductwork and packages.
- Designing for high wind loadings (circa 100mph) on exposed ductwork and packages.
- Designing for explosion blast overpressure.
- Integrating non-standard electrical requirements for packages.
- Including equipment for regulating humidity of the air supply to avoid sore throats and dry eyes from low relative humidity.

Completing layouts

During detail design, any remaining and less serious equipment or ductwork clashes will be fully resolved and ductwork fabrication drawings produced. HVAC ductwork and equipment supports will be designed by structural taking into account any wind and explosion overpressure loads.

Interfaces with other disciplines

Electrical, instrument, piping and structures interfaces will now be fully detailed to allow E&I equipment procurement, piping, and structural fabrication.

HVAC Engineering Deliverables

HVAC produces the following design deliverables:

- HVAC Philosophy explaining the system design.
- Ducting and Instrumentation Diagrams. These are similar in concept to a P&ID but shows the HVAC equipment, ductwork, and packages schematically.
- Duct Flow Diagrams That show how much air is required to flow along each duct.
- Equipment Specifications that allow for procurement of packages.
- Sizing calculations.
- General arrangements and layouts of equipment and ductwork.

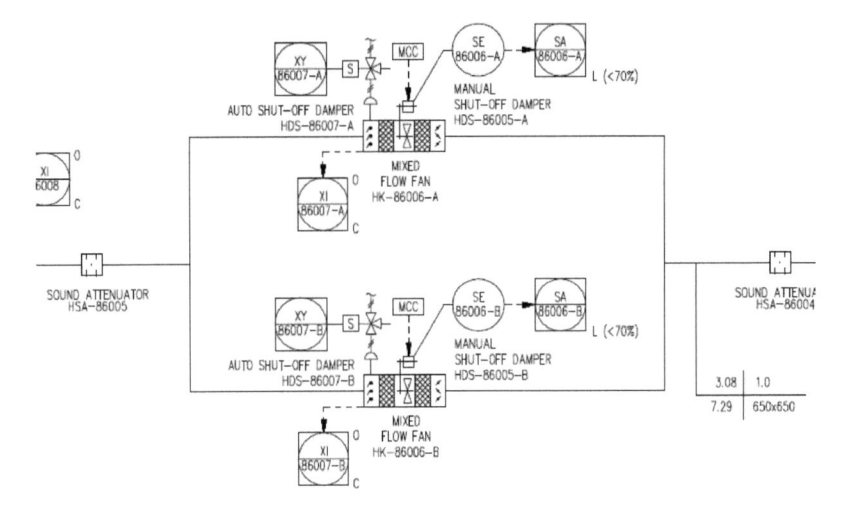

Figure 27.8 HVAC D&ID excerpt.

HVAC Equipment

Whilst there are some multi-equipment manufacturers of HVAC equipment, most manufacturers specialise in individual elements. The typical approach is for the HVAC engineer to specify individual pieces of equipment and to integrate them into the overall HVAC system.

The main long lead (typically 10 to 26 weeks for ATEX rated) HVAC items are:

- Fans.
- Air handling units.
- Split air conditioning units.
- Marine humidifiers.

HVAC package management

HVAC manage each package throughout the design, and manufacturing and testing attending supplier's premises regularly to approve the design and witness testing.

Construction and Commissioning

HVAC will provide support to the construction and commissioning teams assisting in resolving any HVAC-related issues.

HVAC system balancing

HVAC system balancing is the most difficult part of the commissioning process. Flowrates and pressures must be set to all areas to ensure correct air flow in each duct, as well as the correct pressure in each area. Once the system has been balanced, heating, cooling, and pressure controls can be set.

It is critical that the pressure differential between the module and exterior or adjoining modules is within range. If differentials are too high, external doors will be either too hard to open or will fly open, both risking injury. There are many instances of installed offshore HVAC systems where commissioning has not been carried out properly and the HVAC system is not operating correctly.

Routes to Becoming an HVAC Engineer

There are principally two routes to becoming an HVAC Engineer in the oil and gas industry. Most engineers will either have a degree in Mechanical Engineering or Building Services Engineering. The skill sets of these types of engineers are highly transferable and it is commonplace that they have worked in other industries, such as construction, nuclear, pharmaceutical, and shipbuilding.

HVAC Summary

HVAC is not normally a large discipline within a project, but it is vital to each area in ensuring correctly-designed ventilation. There are usually multiple design criteria that should be considered before selecting the appropriate ventilation rates, recalculating the heating and cooling loads to ensure all scenarios are met. Non-HVAC members of the design team are frequently surprised at the complexity of HVAC design and the initial "wind and tin" assumption soon becomes "You need to do all of that?".

CHAPTER 28

Metallurgy

What is the source of the most maintenance cost in an onshore or offshore facility? The answer: corrosion of metals. To put this into perspective, in 2016 NACE (National Association of Corrosion Engineers) estimated the global cost of corrosion to be US$2.5 trillion, equivalent to roughly 3.4 percent of the global Gross Domestic Product [Ref 17]. In a similar vein, pipework leaks in the leg of a North Sea platform caused by corrosion led to evacuation and shutdown of numerous fields with consequential losses of approximately £500million.

Fig 28.1 Metallurgy Mind Map.

There are numerous ferrous and non-ferrous metals used in the construction of process facilities. Corrosion occurs due to external environmental factors such as salt-laden air, and internally due to corrosive and erosive process fluids.

It is the job of the metallurgist to assist in specifying the right materials for a particular application. Specifying the right material involves trade-offs against cost, availability, weldability, etc.

This chapter will take you through the metallurgy discipline and how they interface with other disciplines.

PETER F CRANSTON

Areas Covered in Metallurgy

The following areas are covered by metallurgy:

- Materials selection and assurance.
- Materials welding and assurance.
- Integrity management input.
- Failure investigation.
- Coatings specification.
- Non-metallics.

Materials Selection

Design phase

During the design phase the key deliverable that is produced by metallurgy is the materials selection report. This allows the other disciplines to accurately specify the metals to be used in pressure vessels, structures, piping systems, valves, etc.

Where practical material options exist they are presented in the material selection report to provide greater flexibility to suit their manufacturing capabilities. For example, carbon steel with an internal weld overlay of a corrosion resistant material may be equally acceptable for a pressure vessel as a vessel manufactured entirely from a corrosion resistant material.

For brownfield projects, there is usually also a requirement for reporting an assessment of existing materials to verify the suitability of an existing plant for new process fluids or conditions. This can lead to material upgrades or replacements.

Input information

The input information required by the metallurgist includes:

- Process data.
- Information about existing materials.
- Design life.
- Client materials specifications.

Materials selection

Materials selection starts during FEED, often with the client specifying the types of materials to be used for certain applications. An initial check would normally be made to verify the suitability of the client's preferences. Note that materials specifications, particularly by Tier 1 operators, can be very tight and hard to meet. There is now a trend to move away from onerous and expensive client-specific requests and instead use industry standards and specifications because recent revisions of standards are reaching closer alignment with offshore requirements and expectations.

Materials specification considerations and output to other disciplines

The metallurgist evaluates all the required manufacturing routes, restrictions, and properties of the required materials, including toughness, ductility, percent composition of other metals and elements, weldability, non-destructive testing, inspection requirements, and so on. At the end of FEED, a preliminary materials specification is created and issued to the other engineering disciplines and used to establish the availability and lead time of materials. Some materials are on long leads with production involving steel rolling mills and long-term production schedules. There can also be market shortages when other projects procure the needed materials, leaving a supply "drought".

Exotic and high criticality materials considerations

The first step in procurement of structural or piping materials is to contact suppliers and advising on specifications and required quantities.

High criticality structural steelwork and materials with potential fabrication issues generally have the most onerous specification requirements, so these should be procured as early as possible in the timeline. Suppliers will respond on their availability and will provide actual materials certifications for evaluation by the metallurgist.

Supplier Technical Assurance

Vendor document review

Typically, available materials do not match up exactly with the required specifications, and one of the responsibilities of the metallurgist is to review the supplier's available specification in detail to determine whether this can be used for the application. To provide assurance that any non-fully compliant materials can be used, the metallurgist will often request additional destructive and non-destructive testing.

Destructive testing

Destructive testing involves taking samples of the proposed material and subjecting it to various hardness tests, tensile tests, corrosion tests and others.

Non-destructive testing

Non-destructive testing, especially for primary steelwork, may involve 100% ultrasonic testing of particular materials to ensure that it is free from laminations and discontinuities greater than a certain size.

It is always preferable that deviations to specification are addressed prior to fabrication, but there are many instances where metallurgists are required to qualify materials after fabrication has commenced. This may be due to poor project management or client schedules pressuring commencement of fabrication prior to material qualification.

Code and standard compliance

A large part of the metallurgist's workload is to support suppliers and subcontractors in ensuring that their proposed materials comply with the relevant national codes, standards, and project specifications. Most suppliers do not have the engineering contractor's level of expertise with materials and require support throughout the scope of supplying materials.

Equipment and package suppliers will submit materials and testing certification during their design and manufacturing processes that will be reviewed by the project metallurgist, and a decision will be taken whether it is acceptable for the manufacturer to proceed.

Specification non-compliance

When a deviation is requested, the metallurgist will decide on the outcome, considering the particular fluids and operating conditions to determine acceptance. If there is a major deviation or compliance issue that has a significant impact on the project, it will be referred to the client's technical authority for a final decision.

Welding

Most offshore metallic structures, vessels, equipment packages, and pipework involve welding during fabrication and construction onshore or offshore.

Welding involves the following steps:

- Weld type specification.
- Welding Procedure Specification (WPS).
- Welding procedure qualification.
- Welder coding.

Weld type specification

The engineer/designer specifies the type of weld on the fabrication drawing.

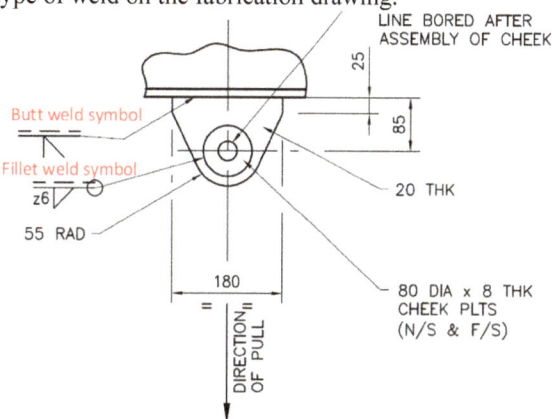

Fig 28.2 Padeye fabrication drawing showing welding symbols.

Welding Procedure Specification (WPS)

The fabricator reviews the fabrication drawing and determines if suitable welding procedure specifications are available. A welding procedure specification is a document describing welding parameters with direction for the welder or welding equipment operators in making sound and quality production welds per code requirements. The document contains weld preparation details, filler material to be used, the current and voltage, travel speed and heat input ranges, number and types of passes, and preheat and post-heat temperatures. If the fabricator does not have a WPS, they will generate one for review and approval. The process to develop a WPS can take many weeks or months and may require adjustments and multiple iterations if qualification fails.

Weld Parameters:

Run No	Welding Process	Filler Make	Trade Name	Dia (mm)	Spec	Grade	AC/DC	Amps (A) (±10%)	Volts (V) (±10%)	ROL (mm) (±10%)	Heat Input (KJ/mm)² (±25%)
1	111	Filarc	88S	2.5	AWS A5.5	E8016-G	DC-Ve	63-78	17-22	36-44	1.3-2.3
2	111	Filarc	88S	3.2	AWS A5.5	E8016-G	DC+Ve	86-106	22-28	79-97	0.9-1.6
Fill	111	Filarc	88S	3.2	AWS A5.5	E8016-G	DC+Ve	98-121	21-27	83-102	1.2-2.1
Fill	111	Filarc	88S	4.0	AWS A5.5	E8016-G	DC+Ve	137-168	20-26	151-185	1.1-2.0
Cap	111	Filarc	88S	3.2	AWS A5.5	E8016-G	DC+Ve	98-121	21-27	118-145	0.8-1.4

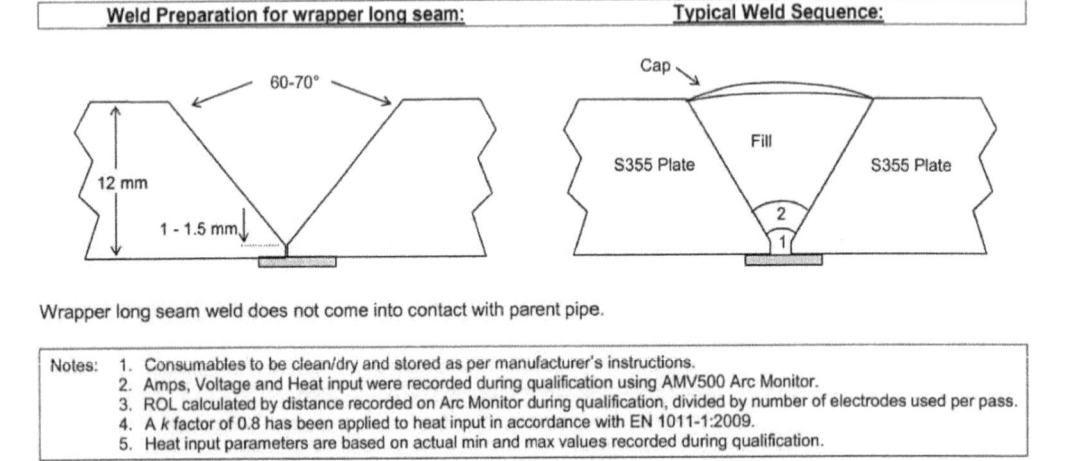

Weld Preparation for wrapper long seam: (60-70°, 12 mm, 1 - 1.5 mm)

Typical Weld Sequence: Cap, Fill, S355 Plate, S355 Plate

Wrapper long seam weld does not come into contact with parent pipe.

Notes:
1. Consumables to be clean/dry and stored as per manufacturer's instructions.
2. Amps, Voltage and Heat input were recorded during qualification using AMV500 Arc Monitor.
3. ROL calculated by distance recorded on Arc Monitor during qualification, divided by number of electrodes used per pass.
4. A *k* factor of 0.8 has been applied to heat input in accordance with EN 1011-1:2009.
5. Heat input parameters are based on actual min and max values recorded during qualification.

Fig 28.3 An example of a WPS.

Weld procedure qualification

A weld procedure qualification will be carried out if a suitable, previously qualified procedure is not available and rigorous destructive testing is needed to ensure that the weld meets the specification requirements.

Welder coding

Finally, the welders carrying out the qualified weld will be examined to demonstrate that they can perform this particular weld.

Once the procedure is completed, the fabrication weld can be carried out. Welders will be monitored to ensure the welding parameters stated on the previously qualified weld procedure are observed, ensuring that satisfactory mechanical properties are achieved in production welds. Production welds are inspected by various techniques to confirm integrity and is used to the extent needed for the application and criticality.

Integrity Management

Metallurgists will provide input into corrosion inspection and management strategies and interface with inspection Contractors, advising on where corrosion and erosion is expected to occur, with the likely rates of corrosion and erosion

Cost of corrosion: 2" piping example.

Metallurgists often provide input and recommendations into inspection and maintenance strategies, advising inspection frequencies as well as when piping systems are likely to fail. Some years ago, metallurgy advised of the risk of failure of a 2" rundown line located in the concrete leg of a large production facility, which was also the main pumping station for numerous interconnected fields.

The client elected not to replace this corroding line as recommended, resulting in failure of the line and flooding of the concrete leg, causing a major platform integrity risk and shutdown of both the platform and pumping station. It is estimated that the lost production cost was circa $700 million. All because of the failure of a 2" run down line!

Failure Investigation

Metallurgists can be expected to be involved in reviewing failures in operating plants to determine failure cause and remedial action. Failures can occur as cracks in welds, pinhole leaks in pipes and vessels, or through wall corrosion or erosion. The metallurgist will then liaise with specialist laboratories and test houses to identify the cause of failure, evaluate the root cause, determine repair procedures, and mitigate impacts in the future.

Pitting corrosion of 316 stainless steel tubes at ambient temperature.

General corrosion of Carbon steel pipe.

Figure 28.4 Examples of corrosion.

Coatings

Painting and galvanising systems fall under the scope of the metallurgist who will ensure that the appropriate coatings are applied to structure and process systems. Given the maintenance cost of corrosion, it is important that they are long lasting and correctly applied.

Non-Metallics

Metallurgists will also be called on to offer advice on correct specification of non-metallics, usually elastomers that are used for seals on valves and rotating equipment.

Cathodic Protection and Dissimilar Metals

Cathodic protection can be used in wetted systems, usually subsea to provide protection to the parent materials by attaching a more electrochemically reactive sacrificial anode. In simple terms, the sacrificial anode corrodes in preference to the parent metal (the cathode). Metallurgists assist with the material specifications and calculations for particular configurations and geometries.

Metallurgists will also screen designs to ensure that there will be no preferential corrosion in assemblies of different materials and will either recommend insulation kits with insulated flanges or alternative materials if there is a potential threat.

Corrosion Under Insulation

Corrosion under insulation(CUI) is a severe form of localised external corrosion that occurs in insulated carbon and low alloy steel equipment. This form of corrosion occurs when water is absorbed by or collected inside the insulation, causing the equipment to corrode as the exposure to water and oxygen is accelerated by higher temperatures. This contamination occurs when the weather-proofing insulation breaks down and allows water to get in. Since it is unseen, the corrosion can continue unchecked. Historical data shows that about 60% of pipe leaks are caused by CUI. The risk of CUI can be minimised by using high integrity insulation sealing systems and ensuring that they are correctly applied.

Routes to Becoming a Metallurgist

The two main routes to becoming a metallurgist are as a materials or corrosion engineer, and as a welding engineer. They typically work simultaneously across multiple projects.

Metallurgy Summary

Many expensive failures and high corrosion rates in offshore equipment and structures can be traced back to poorly specified materials and lack of QA/QC throughout the manufacturing or fabrication processes. The costs of rectification when failures occur offshore are usually orders of magnitude greater than the preventative costs of correct initial specification by the metallurgists. The only caveat to this is to avoid expensive 'gold plating' and unjustified over-specification on either safety of economic grounds.

The importance of receiving final and correct process data before starting detail design cannot be over emphasized. In far too many cases, enormous over-conservatisms are added which increase material selection to unnecessarily costly grades.

"The Metallurgist has been compared with the man who clears out the blocked drains: He is either desperately needed or not wanted."

CHAPTER 29

QA/QC

Quality Assurance (QA) and Quality Control (QC) are the important yet mostly unseen service within a project. However, without QA/QC, a project can fail when materials and equipment do not comply with the specified requirements, often resulting in significant rework costs, client contractual claims, and reputational impact. This chapter will explain the difference between QA and QC and then look into their work processes in more detail.

Quality Control

- PCNs
- Concessions
- External audits
- Inspection planning
- Surveillance & witnessing
- Inspection test plans
- Kick-offs
- Main interfaces

Quality Assurance

- Project QMS
- QA audits
- Quality matrix
- Quality alerts
- Action Request (AR)/ (NCR) System
- Cost of quality
- QA documentation responsibilities

Quality Assurance/ Quality Control

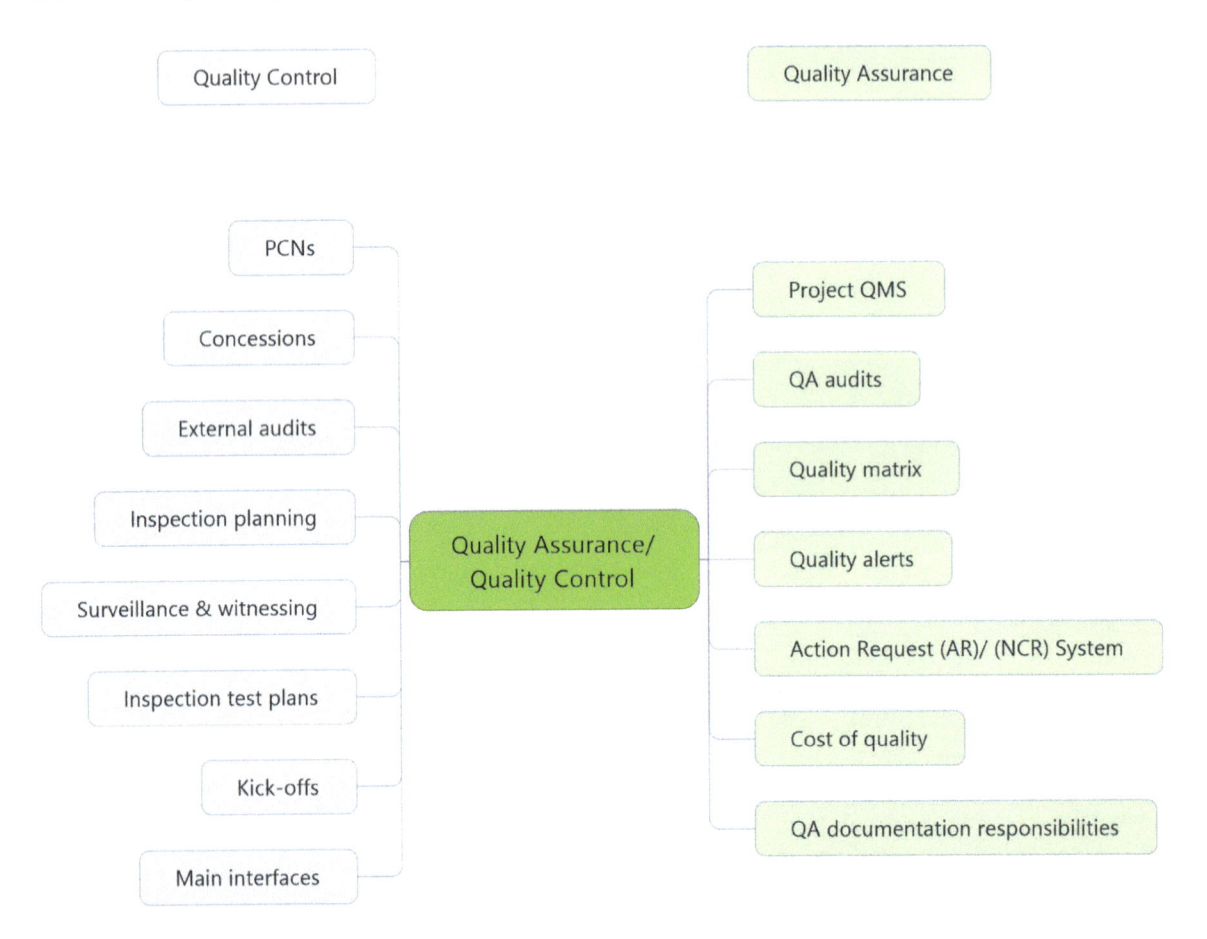

Figure 29.1 QA/QC Mind Map.

QA and QC refer to ways of ensuring the quality of engineering, project deliverables, equipment, and fabrications.

Quality Assurance (QA)

Quality assurance is a method for preventing mistakes and defects in manufactured and fabricated products, avoiding problems when delivering solutions or services to customers. Quality Assurance is all about prevention and includes two principles: "Fit for purpose" (i.e. the product should be suitable for the intended purpose) and "right the first time" (i.e. mistakes should be eliminated).

The Project Management Plan and Quality Management System

Within a project's Project Management Plan, (PMP), there will be a Quality Management System (QMS) that is a collection of processes and procedures focused on consistently meeting customer requirements and enhancing products and services to meet their satisfaction. These documents will be based on overall organisation-, corporate-, and project-specific processes and procedures.

Quality Assurance Audits

Many people are intimidated by audits, believing they are for identifying lacks. However, I view audits more positively because audits ensure that you are complying with project work procedures and seek constructive feedback on problem and improvement areas in the process. Audits are an opportunity to adjust your working practices to fully align with the organisation's procedures and provide feedback on where improvements can be made.

Audits can be split into the following four main types:

- Risk: Aligning with project risk register.
- Client: Carried out by the client or their representative.
- Project phase: Carried out as part of concept, FEED, detail design, etc.
- Trend audits: Review of statistical information to identify trends and remedial actions.

Risk-based audit approach

It is recommended that the QA audit identification is based on the project risk register, with audits scheduled for processes associated with the higher risk activities i.e. there should be a direct correlation between identified risks and the audit schedule. This is in line with the requirements of ISO9001 (the International Standard for Quality Management Systems).

The output of an audit is normally a list of agreed corrective and process improvement actions. These will be followed up on to ensure actions are completed, and if required, the procedure will be revised and reissued as part of a continuous improvement process.

Senior management review audit schedule

QA will meet at key points in the project, typically twice a year, with the senior project team for a management QA review. Based on company requirements and team experience, they will agree on the project QMS, objectives, and schedule of audits that will maximise the assurance provided in the project.

	Audit Title	Project	Audit Lead	Jan	Feb	Mar	Apr	May	Jun	Jul	Aug	Sep	Oct	Nov	Dec
INTERNAL	Estimating		QA Auditor							■	■				
	Invoicing		QA Auditor								■				
	Requisitioning and Procurement of Materials / Expediting		QA Auditor		■										
	Document Management		QA Auditor	■											
	Project Change Control / Deviation and Concessions		QA Auditor				■								
	Project Close Out		QA Auditor										■		
	Workpack Design and Constructability Review		QA Auditor									■			
	PAAF Process		QA Auditor					■							
	Hazard and Risk Management		Engineering									■	■		
	Technical Bid Evaluation		Engineering												
	Design Verification Activities including Document Checking and		Engineering					■							
	Technical Audit		Engineering							■					
	Project Compliance with the Pressure Equipment Directive		Engineering												
EXTERNAL	Supplier A		QA Auditor												
	Supplier B		QA Auditor						■						
	Supplier C		QA Auditor										■		
	Supplier D		QA Auditor												
	Supplier E		QA Auditor											■	

Legend: Audit Planned / Audit Complete / Audit Completed Late / Audit Overdue

Approved: / Date issued: / Revision:

Figure 29.2 Sample audit schedule.

Customer satisfaction surveys

One of the additional roles of QA is to coordinate client satisfaction surveys since you are unlikely to get returning customers and repeat business without satisfied clients.

Quality Matrix

One of the key documents produced by QA is the quality matrix, which summarises the key quality activities that are required to be carried out for each project and sets targets for the performance of each of activity. By managing the detail activities, the overall quality performance of the project is improved.

2016 QMS Deliverables	Target	Project A	Project B	Project C
Implementation of QMS activity martix	100%			
Implementation of QIP activity	100%			
Audit completed by scheduled date (including Engineering audits) and report issued within 2 weeks	>90%			
Audit action closure vs target date	>80%			
Non Conforming Material Report and CAPA closure vs target date	>80%			
Customer Complain Closure v target date	100%			
Procedure review completion by scheduled date	>80%			
Vendor Kick Off Meeting / Pre Inspection Meeting process to be standardised. Using standard KO meeting template	100%			

Fig 29.3 Example excerpt of a quality matrix.

Quality Alerts

Quality alerts are normally industry-wide alerts that advise issues with a specific manufacturer's equipment and are communicated internally to the required discipline engineers for ensuring that equipment is not used and that controls are placed to assure safety and continued operation.

Action Request and NCR System

A non-conformance report (NCR) means that something has gone wrong or a problem has occurred and needs to be addressed. Non-conformances are addressed with corrective actions. Each project QA team will maintain an action request and non-conformance system. For audits the following key items will be recorded:

- Results of investigations into quality issues.
- Non-conformances with procedures or specification requirements.
- Opportunities for improvement.
- Positive findings during audits.

Cost of Quality

The following lists both the costs of confirming to quality and the costs of non-compliance. The cost of quality conformance should always be less than the costs of non-conformance i.e. quality should be a positive investment.

Costs of conformance

- Quality Training.
- Studies.
- Surveys.
- Efforts to ensure everyone understands processes and procedures.

Costs of non-conformance

- Rework.
- Scrap.
- Warranty claims.
- Reputation and lost business.

Gold plating

Gold plating refers to providing equipment of documentation above that was specified and agreed upon with the client. Gold plating can arise when an engineer wishes to create the best possible solution without understanding the impact on project profitability.

The exception to this is when the client perceives added value, such as if an upgrade to a material specification, which results in additional short-term costs but reduces the lifetime maintenance replacement cost to the client in the future. Such exceptions should be submitted to the client as a variation request for approval, advising the perceived benefits and expected additional client costs.

Documentation QA Responsibilities

Workpacks and requisitions

QA will review, check, and approve construction and commissioning workpacks and requisitions prior to issue in order to ensure that they comply with the list of requirements identified in the procedures. This type of checking calls with a detailed eye ensures that all quality non-compliances are identified and corrected.

Lessons learned / progressive learning

Part of the continuous improvement cycle on projects within organisations is the formal capturing of lessons learned and sharing them widely with the organisation via a shared database. QA will typically chair these sessions and record the learnings.

QA Summary

In summary, QA is all about preventing mistakes or defects from occurring and viewing the associated QA costs as an investment rather than a cost.

Quality Control

Quality Control (QC) is a process by which projects and subcontractors review, monitor, and measure the quality of all factors involved in the production of engineering documentation, equipment, and fabrications:

- **Engineering Processes:** Including work procedures and document management systems.
- **Manufacturing & fabrication**: Includes both manufactured equipment and fabricated pipework and structural steelwork
- **Competence:** such as knowledge, skills, experience, and qualifications.
- **Soft Elements:** such as personnel, integrity, confidence, organisational culture, motivation, team spirit, and quality relationships.

Quality control is all about monitoring and measuring against specific drawings and specifications. Quality Control inspectors are the eyes and ears of the procuring engineers. Most organisations will have their own project QC inspectors but will use external agencies when either the manufacturer or fabrication facility is geographically distant, or when specialist inspection techniques and skills are required.

A key trait of the QC inspector is that the drawings or specifications are their "bibles," meaning that they will not accept any non-compliance or defect. The QC inspector will report back to the engineering team who can evaluate whether to accept a concession on quality.

QC Interfaces

The QC team interfaces with:

- Engineering disciplines.
- Document Control Centre DCC. Responsible for control and distribution of all documents and drawings.
- Suppliers.
- Construction.
- Project Engineers.

Quality Control Activities

Attendance at kick- off meeting

QA/QC will attend both project kick-off meetings internally and at the supplier's premises for specific equipment and fabricated items.

Inspection test plans

An inspection test plan is a document prepared by QC supported by engineering disciplines that identifies each step in the manufacturing and fabrication processes, identifying the level of inspection and documentation required at each stage.

Op. N°.	Supplier Activity	Supplier Quality Activity	Procedure Ref.	Acceptance Criteria	Verifying Document	Personnel Responsible	Supplier	Project	Client	3rd Party
01	Contract Review	Review WorleyParsons PO and approved drawings. Identify & review all relevant Codes, Standards & WorleyParsons Specifications as applicable (i.e. PED, ATEX, CE marking)	Supplier's Procedure	Project Purchase Order	Contract Review evidence	Supplier's personnel as defined in their QMS	H			
Note:	Ensure that all relevant specifications have been thoroughly reviewed and that any additional requirements e.g. testing, concessions, inspections, qualifications etc have been duly considered.									
02	Kick-Off Meeting (If requested by Project)	Ensure that the ITP has been submitted to the Project for review prior to the KOM taking place	Supplier's Procedure	Project specified requirements for KOMs	Minutes of KOM produced by Project	Supplier personnel as relevant to the PO	H	H	W	
Note:	As a minimum, the ITP submitted prior to the KOM shall be a draft version prepared in accordance with these guidelines. The draft ITP will be discussed and agreed upon between the supplier and the Project at the KOM.									
03	Project approval of Supplier documents	Review the Project PO and submit all required documentation as specified in the agreed SDS (i.e. Material certificates, Welding and NDT procedures)	Supplier Starter Pack	Project Purchase Order	Project approved documents (see note below)	Supplier's document control/ Quality function	H	H, R/A		

Fig 29.4 Excerpt of inspection text plan.

Inspection text plans will contain REVIEW (R) points, indicating stages where reviews must be carried out and documentation must be submitted without impeding work progress. More importantly, the text plan will contain HOLD (H) points that indicate where work cannot proceed until the HOLD point has been completed and accepted. And finally, there are WITNESS (W) points that require activities to be witnessed by an inspector or engineer.

An example of a WITNESS point would be the assembly of a piece of equipment containing a pressure vessel that cannot commence until a hydrotest has been witnessed and accepted.

Surveillance and Witnessing

QC inspectors will visit manufacturers and fabricators and carry out the following activities:

- Carry out or review non-destructive testing.
- Witness hydrotesting of pressure-containing items.
- Review material specification certificates.
- Carry out physical dimensional measurements on equipment, fabrications, and painting against drawings.
- Check traceability of documentation for materials.
- Review the manufacturer's documentation.
- Release equipment and fabrications that are compliant with specifications and drawings.

Weld procedures review

Significant effort is expended by Quality Control to ensure that weld procedures are fully compliant with associated non-destructive testing, (NDT); hydrotesting, dye penetrant, radiography, and ultrasonic testing and that the welding record documentation is correct.

Inspection Planning

The QC team will closely interface with projects, planning, and supply chain to determine when to send in inspectors to carry out planned QC activities.

Internal and external inspectors

The QC team will have a pool of internal inspectors that they can use both locally and overseas. They frequently supplant their resources by using third-party inspectors who have the relevant required knowledge and expertise that are often locally-based centres of industry in particular countries. On large projects it is not uncommon for an inspector to be dedicated to a particular fabrication and manufacturing site for many months.

External Audits

QC will carry out external audits of suppliers where the consequences of non-conformance are large or where there is an identified problem with the supplier.

Concessions

A concession request is where the supplier or fabricator cannot achieve the required specification. This may be a material specification, a dimension tolerance, a coating specification, etc. The concession request will be provided to supply chain who will evaluate the request and seek a formal decision by the appropriate discipline engineer on whether to accep tor decline the request. They will then formally record the decision and respond to the supplier. A key part of QC is ensuring that a complete and documented audit trail of all concessions is maintained and issued as part of the manufacturing data book.

Project Changes

As QC inspectors are normally on-site and act as the eyes and ears of the engineering team, they will advise when there is a potential change which may impact quality, cost, or schedule.

Quality Control Summary

QC is concerned with defect identification and rectification, and when correctly implemented ensures that all delivered equipment and fabrications are fit for purpose and compliant with specifications.

"Give them quality. That's the best kind of advertising." - Milton Hersey

Chapter 30

Construction

The construction phase in a project is where all the materials and equipment are shipped either to a module yard or offshore, they are assembled into a completed scope ready for commissioning and transferring to operations.

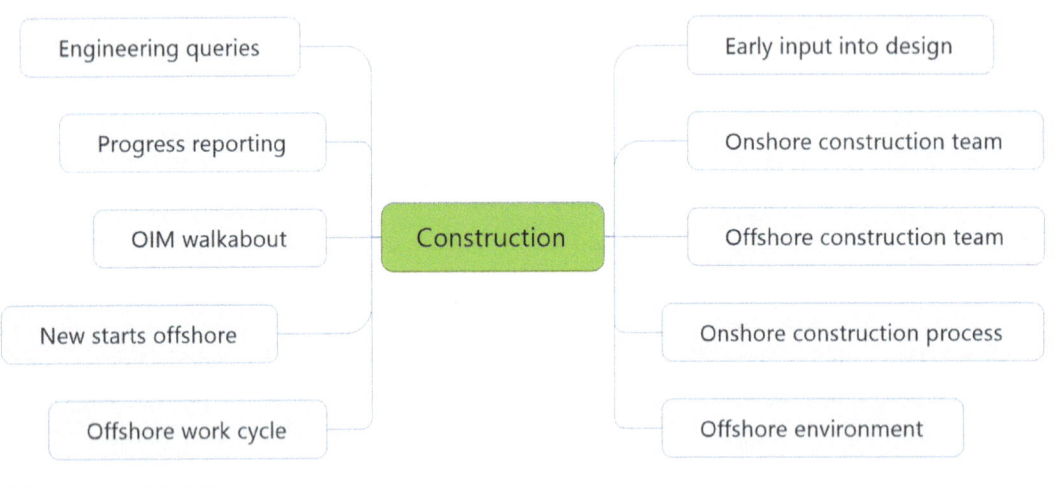

Figure 30.1 Construction Mind Map.

Early Input into Design

It is often assumed that engineering produces the design and that construction becomes involved later on in building he design. However, this would be wrong.

For a project to be successful, it is vital that construction is involved as early as possible, preferably from the conceptual stage. As concepts are produced and refined, it is important to have at least a part-time, experienced construction resource available. Construction should be actively involved in incorporating construction-orientated ideas and solutions to enable the safest and most practical construction.

Working backwards

A technique that I often use is to be armed with the concept(s) when requesting the construction engineer/team to evaluate how they would build it, and advising any design changes that would allow construction complexity or duration to be reduced. This backwards "right to left" approach will be repeated throughout the design and will yield the following benefits:

- Construction build sequence optimised.
- Major construction risks minimised.
- Buy-in to design from construction team.
- Development of positive relationships between construction and engineering.

Tip: Recruit the best and most experienced offshore/site construction engineer to obtain their input, resulting in a payback investment many times over.

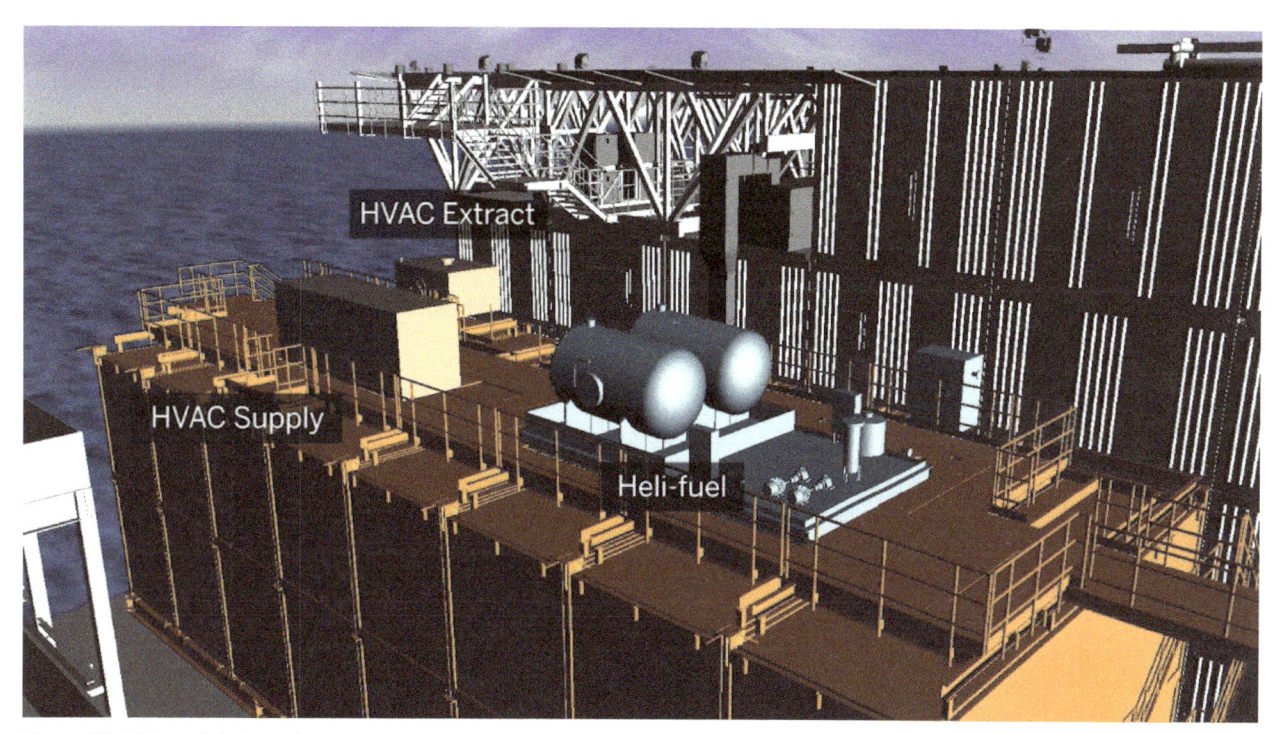

Figure 30.2 3D model view of a new part constructed accommodation block

The above view is a still image from a construction sequence video-generated from a 3-D CAD model. These videos are invaluable when communicating construction sequences to onshore and offshore personnel and well worth the small investment to produce.

Main Construction Functions

There are two main areas managed by construction function:

- Onshore Construction team (sometimes called implementation).
- Offshore/site construction team.

Onshore Construction Team

The onshore construction team will be managed by the onshore construction manager and will have the following responsibilities:

- Formal constructability review of and comment on engineering designs.
- Construction risk assessments and mitigations.
- Compilation of construction workpacks, including jobcards.
- Daily liaison with offshore construction team.
- Preparation of requisitions for construction consumables and equipment.
- Briefing of construction personnel and subcontractors prior to mobilising to the offshore installation.

Offshore Construction Team

The offshore construction team is managed by the offshore construction manager or superintendent who will in turn have dual reporting with the onshore construction manager and the client's offshore installation manager (OIM). This team will have the following responsibilities:

- Demonstrate safety leadership.
- Ensuring adherence to procedures and maintaining a safe work environment.
- Receipt and management of construction materials shipped offshore.
- Receipt and use of construction workpacks.
- Construction of scopes as specified in workpacks.
- Ad-hoc construction support to operations for urgent repairs/minor modifications.
- Accurately reporting progress of construction workscopes.
- Providing support to commissioning teams.
- Daily liaison with onshore construction team.
- Raising engineering or site queries regarding a particular workscope element.

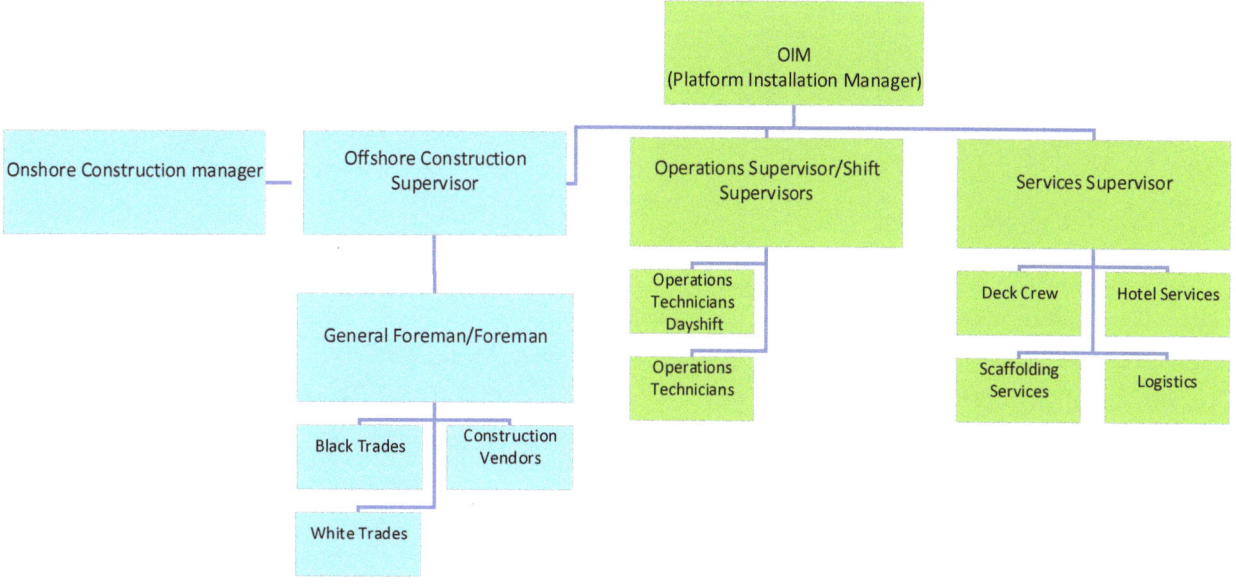

Figure 30.3 Typical offshore organisation.

The Onshore Construction Process

Workpack compilation

The main deliverable of the onshore construction team is "Approved for Construction" workpacks. It is not uncommon for a significant project to have hundreds of individual workpacks. The beauty of the workpack is that it fully aligns with the planning system and WBS, breaking construction down into manageable "bite-sized" pieces.

Construction workpacks are very comprehensive documents that contain critical information, such as:

- Hazard Identification Risk Assessment (HIRA).
- Detailed construction narrative.
- Detailed job cards.
- Drawings list and as built status requirements.
- Bill of Materials (BOM).
- Reference to relevant Handover Certification Pack (HCP).
- Additional information (Data Sheets, Vendor Procedures, etc.).

Hazard Identification and Risk Assessment (HIRA)

These vary in detail from organisation to organisation but all take a structured, documented approach in identifying construction hazards and agreeing on actions and controls to eliminate or reduce risk.

Constructability review

If constructability has been carried out throughout design, the formal constructability review with the draft workpack will be a fairly painless process. This review is the final opportunity to change hopefully minor elements before the pack is issued for construction offshore.

Readiness for offshore construction

Offshore beds are at a premium and there is never enough bedding availability for those requesting beds i.e. operations, drilling, construction, maintenance, hotel services, specialists, etc. It is therefore vital that all possible preparations have been made to ensure maximum efficiency and minimum delays offshore. As the cost of an onshore manhour is significantly less than that of offshore manhours, we can invest hours onshore to save on offshore hours. The following non-exhaustive checklist should be used when checking readiness:

- Workpacks available for construction.
- Specialist tooling and equipment available.
- Materials available offshore or ready to ship.
- Construction permits pre-prepared and available.
- Process equipment isolations identified, agreed on, and available.
- Lifting plans compiled and authorised.
- Bedding requirements identified and agreed upon.
- Third party subcontractors are available.
- Simultaneous offshore operations (SIMOPS) are identified and agreed priorities.
- Final constructability surveys are carried out.
- Pre-investment in scaffolding, lifting aids, etc.
- Crew rotations are compiled and agreed upon.
- Trained and competent personnel are available as per agreed rotation.

Construction readiness gates

Many organisations have a "gated" process for readiness for offshore construction i.e.

- Construction – 6 weeks   Workpack and materials available.
- Construction – 2 weeks   Permits written and available.

The Offshore Environment

The first two things you will notice about the offshore environment is that people are generally calm and do not artificially stress the environment. In addition, it always seems to take an eternity to get anything done. This is partly due to the required strict control of work intended to keep people and the facility safe. Given the high inventory of hydrocarbons on a facility, the consequences and escalation potential of an incident could be catastrophic and result in multiple fatalities.

Offshore trips are typically two to three weeks long with a similar amount of leave. This results in two shifts working on the same scopes. Offshore personnel work a minimum of 12 hours and often longer days.

Tip: When possible, travel to the onshore or offshore construction worksite as often as possible will let you build vital working relationships with the construction team in their own environments. It is amazing what insights into the project you can gain and in order to support and rectify existing and emerging issues. Because you have taken the time to get to know them onsite, they will be more likely to help you when you ask for support.

The Permit to Work System

The "Permit to Work" system refers to management systems that ensure work is done safely and efficiently. These are used offshore and involve procedures to request, review, authorise, document and de-conflict tasks to be carried out.

The permit to work system is a formal system stating exactly what work should be done, where, and when. A responsible, approved, person will assess the work and check safety at each stage. The people doing the job sign the permit to show that they understand the risks and necessary precautions. The permit to work is a core element of integrated safe system of work (ISSOW) that, along with risk assessment and isolation planning, enable reduction of unsafe activities to as low as reasonably practicable (ALARP).

In practical terms, permits are drafted in advance of the construction activity for identifying the scope, hazards and mitigation actions. These are submitted to the permit controller at least one day before the activity is scheduled to start. The permits are then reviewed by the platform management team with any additional controls and precautions added that are then approved and issued for use at the start of the next workshift. The permit must be visibly displayed at the worksite.

The following main types of permits are used by all operators:

Category 1 Hot Work: Work involving an ignition source within an operational area
Category 2 Hot Work: Work involving a potential source of ignition within an operational area
Breaking Containment: Work involving the breaking of containment of systems or equipment containing produced-hydrocarbons or other high-pressure, high temperature or hazardous materials
Cold Work: Any non-routine work within the operational area where it does not fall under Hot Work or Breaking Containment Permits.

It is important to note that the quantity of each type of permit that can be issued will depend on the current platform status, whether producing, shutdown, etc., and the physical locations of the activities with respect to one other.

Finally, on a more sobering note, Occidental Petroleum's Piper Alpha platform was destroyed by explosion and fire after a shift reinstated a system was left partially disassembled by the previous shift. As a result, 167 men died in this incident due to failure to properly communicate permit state at shift handover.

The Offshore Work Cycle

The offshore week and days contain the following main activities:

Weekly lookahead/mission statement

The weekly review usually occurs between onshore and offshore sites via video link/conference phone that will look at the planned scopes for the next one to two weeks reviewing readiness with respect to workpacks, materials, personnel, and permits. This will include both planned scopes and opportune fall-back scopes.

Daily morning call

This is a daily conference call between onshore and offshore construction and often client onshore and offshore representation. In the meetings, any short-term issues will be discussed with requested support from the onshore team. It is common for a pre-meeting call to be held only between the contractor onshore and offshore managers, to discuss potentially sensitive issues and allow a more cohesive structure to the following wider meeting involving the client.

The offshore morning briefing

The construction manager will get the entire team together prior to the start of the shift to brief on any specific issues for the day and to get feedback from the team. The main focus of this meeting will be safety of the team and the facility.

Site activities

After the morning meeting, the teams will don personal protective equipment (PPE) and exit to the various worksites. At the worksite, a toolbox talk will be completed to ensure that all work-party members fully understand the work, the risks and the required precautions.

New Starts

Personnel new to the installation will be "buddied" with an experienced team member, specifically to provide support and ensure their safety as they become familiar with the facility, emergency procedures, and working methods. For new starts who have never been offshore, the buddy will also teach them the required safety culture and how to ensure both their own and their colleagues' safety.

OIM Walkabout

The client offshore installation manager who has overall responsibility for efficiently running the installation and safety of all personnel on board (POB) will conduct a walkabout at least once a day, visiting most areas of the facility where activity is taking place. The focus of the managers is safety, so if required, he/she will stop the activity and call a "time out for safety" if should there be any concerning activities.

Construction Progress Reporting

Work progress is reported as percent complete of each of the jobcards contained in the workpack. Full details of this structure are covered in the planning chapter. One aspect of reporting to be aware of is that construction teams sometimes under-report true progress to ensure that they keep a buffer of hours available for unproductive/lean periods. The solution to continuous under-reporting is to enlist an offshore planner who can visit worksites and establish true progress.

Engineering and Site Queries

When a construction problem is identified offshore, such as a component or section of pipe does not fit, the offshore team do not decide on or implement a solution by themselves. They prepare an engineering/site query, usually supported with photographs advising on the issue and suggesting potential solutions. These are reviewed onshore by the onshore engineering team taking into account all design requirements with a response provided to the offshore site. Offshore staff take the instruction and implementation as requested, and care needs to be taken to ensure that this is rigorously followed as unapproved changes can contribute to an incident causing injury and damage.

Construction Summary

Construction is the "sharp end" of the overall project process where design and fabrication errors become evident. Construction, by its environment, also presents the most risk to people. The key to successful construction is the preparation, organisation, and a continuous focus on the whole team, keeping each other and themselves safe.

CHAPTER 31

Commissioning

As Project Engineer, you and the wider team will have engineered the project, procured all the necessary equipment and materials, and completed offshore construction. You now need to make it work. This is the commissioning phase and is the "Proof of the Pudding" in which any design or construction issues will become evident as the commissioning team "work up" the system(s) before final handover to operations.

This chapter will take you through the logical and structured commissioning process, from onshore documentation and testing to final offshore commissioning.

Figure 31.1 Commissioning Mind Map.

PETER F CRANSTON

Definitions

First, we will present some common definitions that you will encounter in most organisations.

Systems

Each facility comprises of multiple systems, each with a unique identifier, such as Fuel Gas, Potable Water, Oil Export, etc. A system, whose title is usually self-explanatory, can be defined as an interconnected arrangement of plant and equipment that carry out specific functions within the facility.

This allocation of systems and associated numbers is done during the initial facility design stage, allowing all equipment, plant and associated documentation (such as drawings, manuals, etc.) to be assigned to a system. A typical installation will have circa one hundred systems.

Systems typically include:

- WELLHEAD CONTROL PANEL & HPU
- PRODUCTION WELL
- PRODUCED WATER RE-INJECTION WELL
- OIL EXPORT RISER/PIPELINE
- GAS EXPORT RISER/PIPELINE
- CRUDE INLET HEATER
- OIL SEPARATOR
- OIL HEATER
- COALESCER
- COALESCER OIL PUMPS
- COALESCER WATER PUMPS
- OIL EXPORT COOLING
- TEST SEPARATION V-23101
- MAIN OIL EXPORT PUMP
- OIL EXPORT METERING
- GAS COMPRESSION
- GAS DEHYDRATION
- GLYCOL REGENERATION
- SALES GAS FISCAL METERING

The Commissioning Process

The following section is based on information provided by WOOD Commissioning as part of their Go Completions® system.

CHAPTER 31 COMMISSIONING

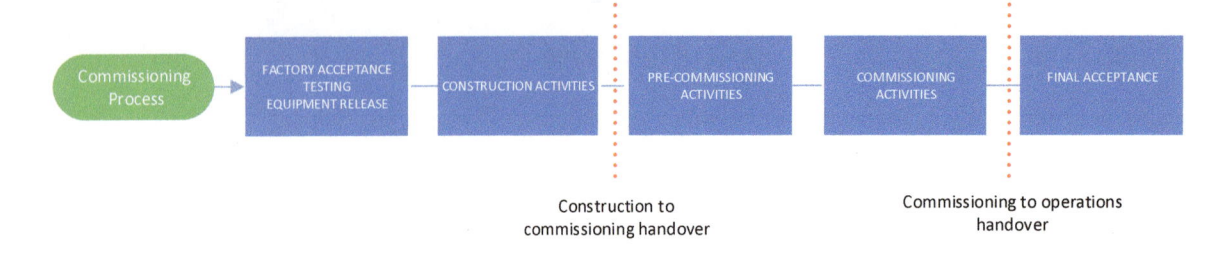

Figure 31.2 Commissioning process.

FAT and equipment release

Commissioning begins onshore with commissioning engineers attending key factory acceptance tests (FATs) in order to familiarize themselves with the equipment and directly obtain any required commissioning clarifications from the suppliers' personnel. Activities that commissioning engineers may participate in include:

- Witness inspection and testing.
- Update and agree outstanding work list.
- Review vendor dossier and manuals.
- Review Inspection Release Notes (IRN).
- Complete verification of equipment functionality.
- Ensure compliance with specifications.

Construction

This covers construction completion, non-energised inspection, and testing carried out by the construction team. Specific tasks include:

- Inspect and check installed items.
- Complete polarity and loop checks.
- Complete hydro-testing.
- Complete cold alignments.
- Carry out flange management.
- Commence preservation of equipment.
- Prepare initial punch lists.

Example of construction completion: A new motor/pump skid has been installed and the pipework and power/instruments have all been hooked up, but the system has not yet rotated under power.

Pre-commissioning

Pre-commissioning, also known as function testing, is when the commissioning team carries out point-to-point wiring checks (such as loop tests) to confirm electrical connections are correct. Once this point-to-point checking is carried out, the commissioning team will power up individual loops and check that the control logic is operating as expected per cause and effects. For the motor/pump skid example above, for instance, the commissioning engineers will "jag" the motor to ensure it rotates in the correct direction under power. It is also common to do 1 to 2 hour (and in some instances, 4 hour) motor runs for recording appropriate operating parameters.

Specific activities during pre-commissioning include:

- Update and agree upon punch lists.
- Clean, flush, and drain pipework and vessels.
- Chemical cleaning.
- Energised equipment checks.
- Carry out Motor no-load runs.
- Final equipment alignment.
- Complete loop and function testing.
- Test control system interface.

Commissioning

The objective of the commissioning stage is to get the system/s to a state ready for start-up by operations. If pre-commissioning procedures and checks have been rigorously prepared and implemented, this should be relatively trouble-free. At this point, any design or construction issues will also become evident and have to be addressed.

Activities during the commissioning stage includes:

- Update/ and agree upon punch list.
- Conduct site integration testing.
- Check cause and effects.
- Perform leak testing.
- Complete media fills.
- Close outstanding technical queries.
- Pre-start-up audit.

Start-up and provisional operations acceptance

This is the "fun" bit where the system will be powered up, and hydrocarbons or other fluids/gases are introduced. Operations are responsible for starting up the equipment and introducing hydrocarbons etc. and they will be supported by the commissioning team during this period who will assist with any necessary adjustments or emergent issues. Operations will take provisional acceptance of the system at this stage.

Activities include:

- Reliability run tests.
- Conduct live ESD testing.
- Perform capacity tests.
- Complete the punch lists.
- Complete "as built" surveys.
- Make "red line" drawings available.

Performance verification and final acceptance

Operations support and final handover

Upon completion of provisional acceptance, there will be a period of weeks or months within which performance will be monitored to ensure that the system is sustainable and that and any emergent issues are addressed. Commissioning engineers may even stay onboard during this period to provide ongoing support and expertise to operations.

Certificates supporting each stage of completion

Accurate certification is required throughout the construction and commissioning stages as well as for the start-up process to ensure that the system, equipment, plants, and processes function safely and as intended. Unfortunately, it is not unknown when pre-commissioning to find gloves, tools, welding rods inside the process equipment. The structured fully-certified approach throughout each stage of commissioning ensures that all the required activities and checks have been completed and are fully documented.

Certification documentation, titles, and coding vary from proprietary system to system, although the following main principles apply to all systems:

Construction certification dossiers are normally completed when the construction workpacks are initially issued and aligned with the proposed construction sequence. These sequenced dossiers are then aligned with the commissioning procedures used to carry out mechanical completion for each system. Once mechanically complete preparations are made for start-up and handover to operations. Completions certification covers the following five main steps:

- Construction completion.
- Mechanical completion.
- Ready for start-up.
- Initial handover.
- Final handover.

Construction completion: As each sub-system is physically constructed and inspected, a certificate documenting each sub-system will be produced. When all the sub-systems have been fully constructed, the commissioning engineers will produce a Construction Complete Certificate (CCC) for that complete system.

Mechanical completion: After construction, the commissioning team will take ownership of the CCC-certified system and then carry out pre-commissioning resulting in a Mechanical Completion Certificate (MCC) for that system.

Ready for start-up: After obtaining the required number of MCCs, the commission team will carry out the commissioning activities resulting in a ready for "start-up certificate", (RSU), for that group of systems.

Initial handover: At this point commissioning will hand the systems over to operations, who will then introduce hydrocarbons. When successfully completed, an interim handover certificate (IHC) will be issued.

Final handover: After the contractually-agreed performance for the given time period and addressing any emergent issues, the Final Handover Certificate (FHC) will be issued.

Completions Management Systems

The most well-known approach for commissioning in the Oil and Gas industry is the Guidance and Certification (GoC) system, which was initially developed by BP and QEDI in the 1980s and has since been used in the wider industry. Whilst the aforementioned commissioning process refers to the WOOD/QEDI methodology, there are others available, including individual operators using their own proprietary systems.

Regardless of the methodology used, however, they all share a similar structured hierarchy. Frequently-used commissioning management systems include ORBIT™, PIMS, WinPCS™ and GoCompletions®.

When certification is completed, there will normally be an accompanying punchlist(s) for items that have not yet been completed, such as painting touch-up or pipework insulation. Punchlist items are recorded as either Category 1 (significant issues) or Category 2 (minor issues) items; the process can proceed with Category 2 items, but the commissioning team will not accept a system with any Category 1 items and will normally put a limit on quantity of Category 2, (minor issues) that they will accept.

Inputs to the Commissioning Process

Commissioning will use the following key inputs for documentation generation:

- P&IDs.
- Cause and effects.
- HAZID.
- HAZOP.
- Equipment types and information.
- Inter-system interactions.
- Vendor information.

The lead commissioning engineer and his/her team will then review the P&ID and cause and effects, and then mark-up (with colour) system limits, also known as Battery Limits. These colour-coded limits will delineate the scope to be commissioned. In the excerpt below, you can see two-colour coded systems, in green and yellow. Note the yellow and red flags that are used to indicate the limits of each system; they are usually located on one side of a valve.

CHAPTER 31 COMMISSIONING

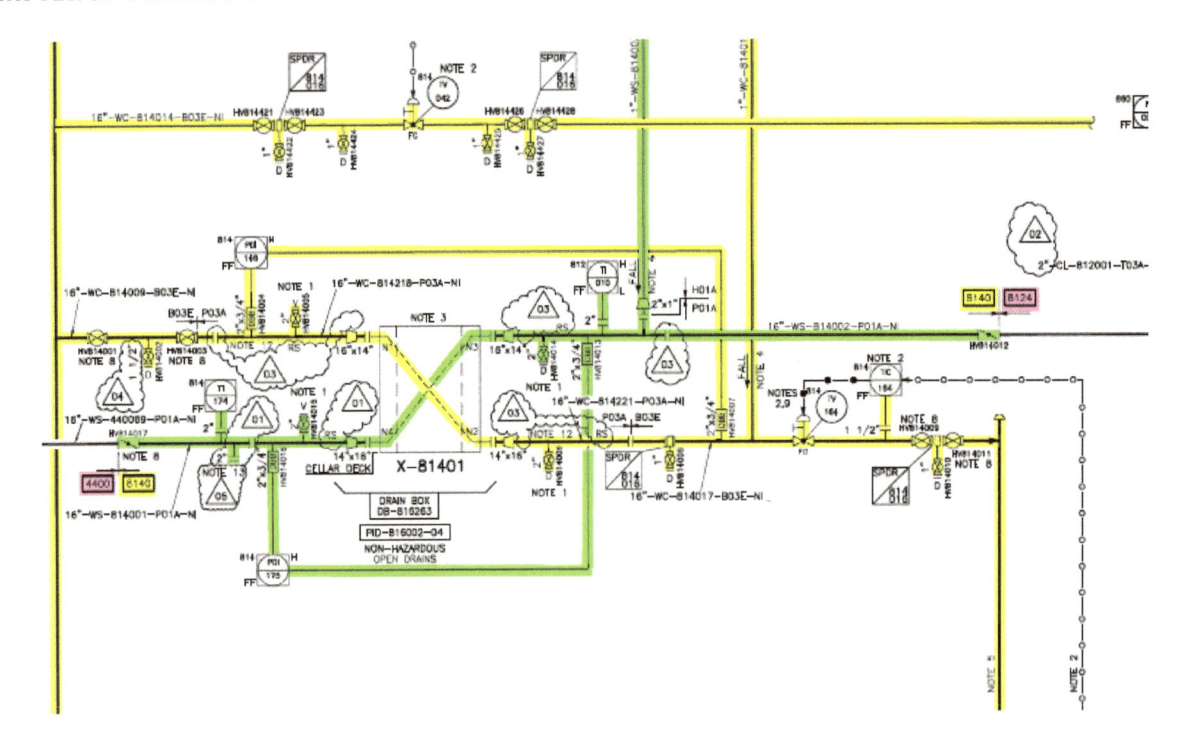

Figure 31.3 Example of system limits.

Equipment information, including specifications, schematics, and general arrangements, will allow the commissioning team to identify the requirements for commissioning each piece of equipment and prepare subcontractor work scopes. Typically, the commissioning lead engineers will join the project in detail design in order to prepare commissioning plans and documentation in good time for onshore testing and offshore commissioning.

Commissioning Preparation Outputs

The commissioning team prepares a wide range of documentation prior to mobilising offshore, including:

- Marked-up system limits on P&IDs and cause and effects.
- Commissioning workpacks.
- Vendor work scopes and vendor commissioning procedures.
- System and subsystem commissioning sequence/logic (see Figure 31.4).
- Integrated schedule.
- SIMOPS identification.
- Offshore permits.

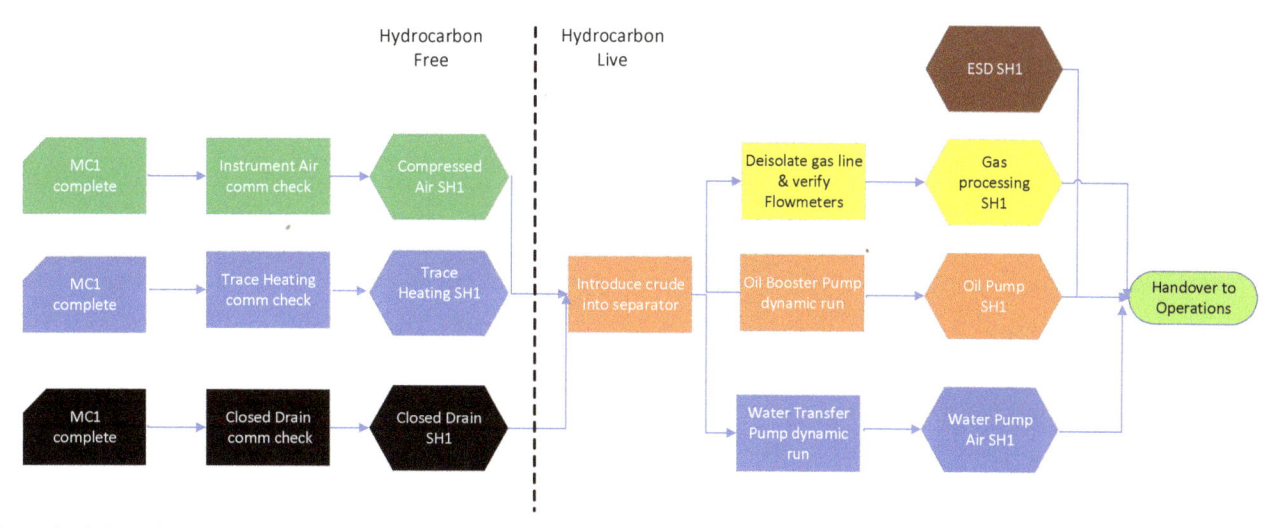

Figure 31.4 Commissioning sequence example.

Figure 31.4 shows a simplified, fictional example of how mechanical completion (MC1) of each system moves through commissioning checks before hydrocarbons are introduced, start-up with hydrocarbons and actions to complete commissioning and final handover to operations. Colour-coding can be used to identify the different systems. These flowcharts are useful for briefing teams involved in commissioning offshore and can be easily marked up to clearly display progress.

Onshore Commissioning Activities

Involvement early in the project

It is important that the commissioning team is involved early in the project when the project Work Breakdown Structure (WBS) is being generated to define systems and limits and thus assist the smooth flow from construction through commissioning, and handover to operations. By considering how the plant will be commissioned, the optimum construction sequence can be developed, which in turn will guide how workpacks and engineering deliverables are produced.

Onshore pre-commissioning

Many systems can be commissioned onshore, which has a number of benefits:

- Testing can be completed in the vendors' facilities, where all the required resources are available.
- Minimising the offshore commissioning durations and impact on precious beds.
- More cost effective to do onshore.

Procedure development

The commissioning team will develop step-by-step procedures for each system and depending on complexity, the documents can be very large, running to hundreds of pages long. The team will also liaise with the engineering team and vendors to ensure that vendor procedures for offshore commissioning are produced and fit for purpose.

CHAPTER 31 COMMISSIONING

Completions management system verification

Completions management system verification involves checking the documentation associated with each step of the completions process for quality and accuracy of information. The approach is quite rigorous, requiring buy-in from the entire team associated with construction and commissioning, and as such will resulting in an efficient and safe handover of the plant and equipment.

Tip: I recommend that formal CMS training be considered for team members, with at least one audit conducted, to ensure that the completions process is followed correctly.

Factory Acceptance Tests

Prior to accepting an equipment package offshore, such as a pump set, it must be tested onshore and witnessed by the requisitioning engineer and often a commissioning engineer. This allows familiarity to be gained with the equipment and establishes communication with the vendor site engineers who may also mobilise offshore.

Onshore site acceptance tests

Onshore site acceptance tests are similar to FAT but are carried out when the smaller packages are integrated into a larger package/module prior to shipping offshore.

Large scale onshore commissioning

In the 1970s' and 1980s' platforms were constructed from upwards of 25 individual modules that were installed, hooked up, and commissioned offshore. Due to the large quantity, the process was very time and labour intensive, and as such delayed potential of first oil by years. Today, completely integrated topsides of 10,000 tonnes are assembled onshore with numerous fully-tested systems up and running before the topsides even leave the quayside. Once lifted onto the preinstalled jacket offshore commissioning times can be reduced to months.

Operator engagement

The offshore facility control room operators and technicians are ultimately the personnel who will operate and maintain the plant, so they are involved at appropriate stages for familiarisation and providing key feedback. The opportunity to provide feedback allows for changes to be made prior to handover offshore, such as modifying control room graphics for a new piece of equipment for ease of use.

It is very important that the graphics, format colours, alarms and such are in designed to be compatible with the operation team's requirements, including how they may already be used to working with other equipment.

Offshore Commissioning

The commissioning team mobilises offshore once construction is nearing completion. They are always seeking to pull forward construction completion to allow release of systems that they can commission.

Construction and commissioning interfaces

Construction is typically completed as a physical area on a facility whereas commissioning is completed by an identified system that may cover multiple physical areas, as such making planning interfaces more challenging.

Triparty walkdowns and punchlists

Once construction activities are completed, the system is handed over to the commissioning team for review and acceptance. The handover will include a punchlist detailing any items that, for various valid reasons, have not yet been completed. After reviewing the documentation, the commissioning, construction and operations teams will complete a "triparty" walkdown of the system using the marked-up P&ID to satisfy themselves all is as per documentation. Once the walkdown, commissioning will take ownership of that system.

Walk the line & walk it again, and with P&ID's document any discrepancies

Permit preparation and management

As all work is controlled on offshore facilities by a permit system a significant part of the lead commissioning engineer's responsibilities is preparation of permits. Permit preparation involves liaising with the operations and construction teams and specialist vendors to manage the commissioning activities, as well as identifying boundary isolations, piping, and the E&I systems that will be applied by authorised operation's personnel.

Short-term planning

The lead commissioning engineer will attend the daily permit meetings to assist in SIMOPs and defend any challenges concerning the commissioning process.

Competency of personnel

Commissioning personnel who will be working on systems are required to have a number of competencies:
- Trade competencies for their specialism. I.e. instrument technicians.
- Platform competency gained after a number of offshore trips and examination/auditing of competency by platform area authorities.

Commissioning

The commissioning team, which may be up to twenty for small projects and over a hundred for larger projects, will commissioning the facility in line with the approved and risk assessed commissioning procedures.

Onshore and Offshore Discipline Engineer Support

Onshore support will be provided by the onshore engineering team who designed the facility.

Tip: When possible, use the engineers who designed the system to support the offshore commissioning team onsite since they not only have the relevant design expertise but will also gain invaluable feedback on what works well (or not so well) to incorporate into their future projects.

Energising systems

All commissioning engineers have at least one "Oops" moment relating to energising a system. These unplanned activities can have significant safety and cost implications.

CHAPTER 31 COMMISSIONING

Example: When commissioning a 5MW water injection pump costing many millions of dollars, the commissioning engineer followed procedure and ensured that the shaft bearing end thrust monitoring readings met specifications and that the main pump thrust bearing oil feed sight glass was full. When commissioning checks were completed per procedure, the commissioning engineer powered up the 5MW pump set. Within minutes, however, there was a tremendous noise and a large, overloaded dry thrust bearing melted and disintegrated. Fortunately, there were no injuries, although there was significant damage to the equipment and as such a deferment of oil production.

So what happened? The procedure had been followed! Two things actually occurred:

- The end thrust monitor was physically positioned in the wrong place and as such provided incorrect readings. This was both a design and manufacturing error.
- The oil lube feed pipe was blocked and as such the bearing was starved of oil.

I have been in similar situations and experienced the gut-wrenching realisation of what you have done.

Tip: When commissioning, check and double-check everything and assume nothing.

Routes to Becoming a Commissioning Engineer

There is no obvious single route, degree, or apprenticeship in commissioning. All routes tend to be through the disciplines; mechanical, electrical, instrument, process engineering, etc. Typically, being time-served and experienced in a particular discipline lends itself to becoming a commissioning engineer.

Specialist engineers such as gas turbine, rotating equipment, and bespoke equipment tend to be "in house" engineers who are directly engaged in the start-up alignment of equipment. These people are invaluable and will support you (at cost, of course) during the dynamic commissioning phases.

In general, experience is gained in the typical fashion of "see one, do one, teach one," meaning soak up as much information as possible and get yourself involved at every opportunity, including Factory Acceptance Tests and Site Acceptance Tests and allow this to propagate into a site based environment.

Commissioning Summary

The commissioning engineers get to play with the "big bits," but with this comes serious responsibility to ensure that everything is done correctly and kept safe. There are many systems offshore that operate sub-optimally, and in many cases, this is due to partly or poorly commissioned systems motivated by a drive to reach production at all costs.

I think of commissioning as a specialist subproject that needs scoping, planning, resourcing and implementation. Take the time to do it right the first time! Doing it right is always the quickest in the long term.

Check and double-check, or as they say in Finnish, "tarkistaa ja tarkistaa".
- Ekku Raikamo

Glossary

3-D CAD	3-dimensional computer aided design
AC	Alternating current
AFC	Approved for construction
AFD	Approved for design
AFP	Active fire protection
ALARP	As low as reasonably practicable
ASHRAE	American Society of Heating, Refrigerating and Air-Conditioning Engineers
ASME	American Society of Mechanical Engineers
ATEX	Appareils destinés à être utilisés en ATmosphères EXplosives. Describing what equipment is allowed in an environment with an explosive atmosphere
CFD	Computational fluid dynamics
CIBSE	Chartered Institution of Building Services Engineers
CMS	Completions management system
CoSHH	Control of substances hazardous to health
CPM	Critical path method
CRB	Contract review board
CTR	Cost time resource
CUI	Corrosion under insulation
DC	Direct current
EEMUA	Engineering Equipment and Materials Users Association
EER	Escape evacuation and rescue

GLOSSARY

ENVID	Environmental impact identification
ESD	Emergency shutdown
FAT	Factory acceptance test
FEED	Front end engineering design
FPAL	First point assessment limited
FPSO	Floating production storage and offloading
GA	General alarm
GA	General arrangement
HAZID	Hazard identification
HAZOP	Hazard and operability study
HIRA	Hazard identification and risk assessment
HVAC	Heating ventilation and air conditioning
I/O	Input/Output
ICS	Integrated control system
ICSS	Integrated control and safety system
IEC	International electrotechnical commission
IP	Ingress protection
IRPA	Individual risk per annum
ISO	International Organisation for Standardisation
KPI	Key performance indicators
LP	Low pressure
Lux	SI derived unit of illuminance
MAC points	Manual alarm call points
MAH	Major accident hazard
NACE	National Association of Corrosion Engineers, an international, not-for-profit professional organisation for the corrosion control industry
NCR	Non-conformance report
P&ID	Piping and Instrumentation diagram
PA	Personnel announcements
PCN	Project change notification
PFEER	Prevention of Fire and Explosion, and Emergency Response
PFP	Passive fire protection
PLL	Potential loss of life
PO	Purchase order
POB	Personnel on board
PPE	Personal protective equipment
PSR	Procurement status register

PUWER	Provision and Use of Work Equipment Regulations
QMS	Quality management system
QMS	Quality management system
ROS	Required on site
RPE	Responsible person electrical
SAT	Site acceptance test
SCS	Safety control system
SIF	Safety instrumented function
SIL	Safety integrity level
SIMOPS	Simultaneous operations
SIS	Safety instrumented system
SLD	Single line diagram
SSJ	Single source justification
T&Cs	Terms and conditions
TBE	Technical bid evaluation
TEMPSC	Totally Enclosed Motor Propelled Survival Craft
UKCS	United Kingdom continental shelf
WPS	Welding procedure specification

GLOSSARY

PETER F CRANSTON

References

1. The inefficiencies of multitasking. Adam Gorlick. Stanford Report, August 24, 2009 URL: https://news.stanford.edu/news/2009/august24/multitask-research-study-082409.html

2. Modern theory of conflict: Course Hero. URL: https://www.coursehero.com/file/pvagfg/TRADITIONAL-VS-MODERN-VIEWS-OF-CONFLICT-TRADITIONAL-VIEW-MODERN-VIEW-1-Conflict

3. Accepted major sources of project conflict: PMBOK?

4. YAMMER: Wickipedia
   URL: https://en.wikipedia.org/wiki/Yammer

5. McGregors theory of X and Y: Wickipedia.
   URL: https://en.wikipedia.org/wiki/Theory_X_and_Theory_Y

6. Maslows hierarchy of needs
   URL: https://en.wikipedia.org/wiki/Maslow%27s_hierarchy_of_needs

7. Toastmasters International: Global communication and leadership organisation.
   URL: https://www.toastmasters.org/

8. Fayol Five Functions of Management. 1916
   https://www.toolshero.com/management/14-principles-of-management/

9. Mintzberg Ten Managerial Roles. 1990
   https://www.mindtools.com/pages/article/management-roles.htm

10. Five Forms of Power by J French & B Raven 1959
    URL: https://www.toolshero.com/leadership/five-forms-of-power-french-raven/

11. Critical Path method Morgan R. Walker & James E. Kelley Jr.
    URL: https://en.wikipedia.org/wiki/Critical_path_method

12. Definition of Risk Management: Association for Project management Book of knowledge.
    URL: https://www.apm.org.uk/body-of-knowledge/delivery/risk-management/

REFERENCES

13. Alexander Kielland offshore disaster 1980
    URL: https://en.wikipedia.org/wiki/Alexander_L._Kielland_(platform)
14. Piper Alpha Offshore disaster 1988
    URL: https://en.wikipedia.org/wiki/Piper_Alpha

15. Pool Fire heat flux. Health & Safety Executive
    URL: http://www.hse.gov.uk/offshore/strategy/pool.htm

16. Heinrichs Law. Herbert Heinrich
    URL: https://en.wikipedia.org/wiki/Herbert_William_Heinrich

17. nace study estimates global cost of corrosion at $25 trillion annum. NACE International, March 8, 2016
    URL: https://inspectioneering.com/news/2016-03-08/5202/nace-study-estimates-global-cost-of-corrosion-at-25-trillion-ann

Index

3-D model, 166, 167, 169, 170, 186, 189
3-D Model Reviews, 166
3-D walkthrough, 167
9 Bar Blues, 92
Achievement moral boosters, 77
Action Request, 257
Active and passive fire protection, 219
Actuated Instrument Valves, 199
ALARP, 222, 267, 282
Allowable strains, 188
Allowable stresses, 188
Analogous estimating, 104
Anticipated Final Cost, 111
Approved budget, 111
As-Built, 162, 167
Association for Project Management, 25, 141, 144
Audience Engagement, 68
Audit Trail, 139
Award recommendation, 176
Backloaded material, 132
Bagging and tagging, 131
Base project estimate, 103
Batching, 39
Behavioural Safety, 234
Best available technology assessment (BAT), 180
Bid conditioning, 175
Black Start, 205
Blast overpressure studies, 220
blow down, 194, 221
Bodies of Knowledge, 141
Brownfield projects, 186
Cable calculations, 213
Cables, 210, 211, 213
Calculations, 94, 161, 163, 213
Cathodic Protection, 252
Cause and effects, 195, 200, 202, 217, 275
Certification data book, 177
Chairing meetings, 33
Change Control, 48, 107, 134, 135, 138, 139
Change Coordinator, 20, 139
Clarifications, 127, 175
Classes (Types) of Estimates, 103

Client interface, 23
Coatings, 247, 252
Codes and Standards, 242
Cold Work, 267
Commercial bid evaluation, 126
Commissioning, xii, 162, 175, 244, 270, 271, 272, 273, 274, 275, 276, 277, 278, 279, 280
Commissioning Process, 271
Commitment, 111, 115
communication, 278, 286, 295
Communication methods, 24
Competency, 279
Completions management system, 278, 282
Completions Management Systems, 275
Compositions, 155
Concept Select Studies, 155
Conceptual study, 158
Concession requests, 177
Concessions, 261
Condition monitoring, 197
conflict, 44, 51, 52, 53, 54, 55, 56, 57, 62, 84, 85, 267, 286
Conflict, 286
Conflict Management, 51
Constructability, 185, 266
Construction, 48, 150, 161, 183, 214, 223, 244, 259, 262, 263, 264, 265, 266, 268, 269, 272, 274, 278, 279
Contingencies, 120
Contingency, 103
Contract negotiations, 128
Control of hazardous substances, 229
Control valves, 199
Core project management, xiv
Corrosion monitoring, 197
Corrosion Under Insulation, 252
CoSHH, 228, 229, 242, 282
Cost control, 111
Cost Engineer, 112
Cost Management, 109, 112, 115
Cost of corrosion, 251
Cost Reporting, 109, 111
Costs of conformance, 258

Costs of non-conformance, 258
Cranes and Lifting Equipment, 179
Critical Path, 91, 286
critical path method, 91
Criticality assessments, 174
Criticality rating assessments, 189
Cultural Differences, 54
Culture, 71
Deck space management, 132
Delegation, 41, 43, 45, 47, 49, 50
Demotivation, 62
Design codes and design verification, 187
Desk expediting, 130
Destructive testing, 248
Detail Design, 161
Deterministic estimate, 121
Diesel power generation, 205
Dilution, 237, 239
Dissimilar Metals, 252
Distributed control system, 193
Distribution boards and switchboards, 210
Documentation for purchase, 176
Documents, 135, 150, 162, 163, 188
Drawings, 163, 202, 265
Dropped object studies, 223
Electrical control and protection, 210
Electrical Drawings, 213
Electrical Equipment Package Management, 213
electrical generation, 205
electrical power, 204, 205, 206, 210
E-mail triage, 39
e-mails, 29, 37, 39, 41, 86, 139
E-Mails, 41
Emergency shutdown system, 193
Emotional Intelligence, 75, 76
Engineering Disciplines, xiii, xv, 146, 151
Enquiry, 125, 127, 175
ENVID, 161, 282
Equipment data sheets, 159, 160
Equipment requisition, 174
ergonomics, 224
Escape evacuation and rescue analysis, 221
Estimate Challenges, 106
Estimating, 101, 102, 103, 104, 108
Ethical behaviour, 127
Exceptions, 107
Exotic and high criticality materials, 248
Expediting, 130
Expenditure phasing, 111
Expert judgement, 104
External Audits, 261
External change, 137
Facebook, 42
Facilitating, 26
Factory Acceptance Testing, 177, 200
Factory Acceptance Testing (FAT), 177
Factory Acceptance tests, 278

FAT, 177, 199, 200, 213, 272, 278, 283
Fayol, 72, 286
FEED, 135, 158, 159, 160, 161, 162, 167, 168, 186, 188, 216, 217, 218, 240, 241, 247, 248, 255, 283
feedback, xi, 278, 279
field equipment, 192, 198
Field expediting, 130
financial forecasts, 110
Fire & Explosion Analysis, 218
Fire & Gas Detection, 215, 225
Fire and gas system, 193
Flare radiation modelling, 221
flow assurance, 164
Free issue items, 131
Front lining plan, 95
Gas Dispersion Analysis, 219
Gas Stream, 154
Gas turbines, 178, 205
Gold plating, 258
Graphical interface, 196
Greenfield projects, 186
Handling challenging personalities, 33
Hazardous Area, 201, 212, 224
HAZID, 223, 275, 283
HAZOP, 159, 161, 217, 275, 283
Health and Hygiene, 228
heat and mass balance, 158
Heavy lifts, 183
Herbert Heinrich, 231, 287
Hierarchy of control, 229
High voltage, 209
Hire equipment, 132
Hot Work, 267
HP/LP interface register, 161
HSSE Advisor Responsibilities, 234
Human factors, 223, 224
human machine interface, 196
HVAC, 236, 237, 238, 239, 240, 241, 242, 243, 244, 245, 283
HVACApplications, 236
HYSYS, 159, 164
inefficiencies of multitasking, 38, 286
Ingress Protection Ratings, 201, 212
Inherently Safe Design, 215, 216
Initial handover, 274
innovation, 49
Input/Output types, 196
Inspection test plans, 260
Integrity Management, 251
Interface Management, 149
Interface process, 149
Interfaces, 148, 243, 259
Internal change, 137
Intrinsically Safe Systems, 202
IP ratings, 201, 212
ISO 31000, 117
Jet and pool fires, 218

INDEX

Jet fire simulation, 219
KANBAN, 148, 150
Key performance indicators, 95, 283
Kick- off meeting, 176
Lagging indicators, 233
Layouts Management, 165
Leader Roles, 73
Leadership, 22, 76, 229
Leading indicators, 233
Level 1Logic, 90
Licensing, 233
Life-savingequipment, 222
Lighting, 211, 212, 213
Line lists, 162, 163, 167
Load list/schedule, 206
Load shedding, 206
Load types, 207
Locked valve register, 161, 162
Low voltage, 209, 212
MAH Register, 222
Management Reserve, 103
Manifesting, 131
Maslow's Hierarchy of Needs, 59, 60
Material shortages, 133
Material storage, 132
Materials Selection, 247
Materials specification, 248
matrix, 58, 147, 256, 257
McGregor's Theory of X and Y, 59
Mechanical completion, 274
mechanical package, 173, 175, 178, 179
Medium voltage, 209
meetings, 279
Meetings, 26, 27, 28, 36, 41, 70
Metallurgy, 246, 247, 253
Metering systems, 197
mindmap, 270
Mintzberg, 286
Minzberg, 73
Mobile Phones, 42
Monte Carlo simulation, 121, 123
Motivation, 58, 59, 62, 63, 76
Motivators, 60
Motor control centres, 210
Multi-criteria decision analysis, 180
Multiplication of effort, 38
NACE, 287
NCR System, 257
Non-destructive testing, 248
Non-Metallics, 252
norms, 54, 106, 221
obsolescence, 197
Occupational health, 228
Offshore Construction.Team, 264
Offshore Environment, 266
Offshore timewriting, 112
Offshore Work Cycle, 268

Oil and Gas Production Process, 152
Oil Stream, 154
OIM Walkabout, 268
Onshore Construction.Team, 264
Onshore Materials Management, 130
Operability review, 158
Operating philosophies, 159, 161, 163
Operator engagement, 278
Opportunity, 69, 116, 118
organisations, xi, xv, 22, 25, 72, 73, 83, 92, 102, 103, 106, 117, 124, 125, 127, 133, 135, 136, 141, 142, 143, 258, 259, 266, 271
organogram, 147, 148
Organograms, 147
Overheads, 105
overspends, 110, 112, 114
P&IDS, 167
Package Management Process, 174
Parametric modelling, 104
Percentage Capacity Checker, 48
Performance Nerves, 69
Performance standards, 224
Performance Standards, 215, 216, 223
Performance verification, 274
Permit preparation and management, 279
Permit to Work System, 267
Personnel protection, 237, 238
PFDs, 147, 156, 163
Pipe sizing, 159, 163
Pipe stress review, 169
Piper Alpha, 193, 226, 267, 287
Piping, 21, 160, 165, 167, 169, 170, 171, 283
Plan entry criteria, 97
Planning, 31, 48, 89, 92, 93, 94, 95, 97, 99, 100, 261
PMBOK, 286
Pollution prevention, 234
Position Statement, 20, 86, 88
Power, 74, 205, 206, 207, 208, 211, 212, 286
Power Distribution, 208
Power system study, 207
PowerPoint, 65, 66
Pre-commissioning, 272
Pre-inspection meeting, 176
Pre-qualification, 127
presentation, 64, 65, 66, 67, 68, 69, 70, 80, 98
Presentation Skills, 64, 70
Preservation, 131
Pressure vessels, 178, 179
Pressurisation, 237, 238
Primary structures, 184
Probabilistic estimate, 121
Probability/impact assessment table, 118
Process Basis of Design, 158
Process Flow Diagram, 156
Process line list, 159, 160
Procurement, 124, 129, 150, 283
Produced Water, 154

Production profiles, 155
Production specifications, 156
Project Management Institute, xv, 25, 102, 141
Project Management Organisations, 141
Project Psychology, 71, 78
Project Structure, 58
Protection devices, 210
Provision and Use of Work Equipment Regulations, 223, 283
PSR, 129, 130, 133, 283
psychological contract, 71
Purchase Order, 125
Q, 254
QA/QC, 191, 253, 254, 259
Qualifications, 24, 143, 144
Qualitative risk assessment, 119
Quality Alerts, 257
Quality Assurance Audits, 255
Quality Control, 176, 254, 259, 260, 261
Quality Management System, 255
Quality Matrix, 256
Quantitative risk assessment, 221
Ready for start-up, 274
Recommendation to award, 127
Relief valves, 199
Risk, xi, 116, 117, 119, 120, 121, 123, 144, 221, 226, 231, 255, 265, 286
Risk Assessment Management, 229
Risk management, 116, 117, 120, 123
Risk Management, xi, 116, 117, 123, 144, 226, 286
Risking, 100, 107
Rotating equipment, 178, 179
Run-Down Curves, 98
S Curves, 98, 115
Safety Alerts, 232
Safety Case Impact Assessments, 222
Safety conversation, 232
Safety Critical Systems, 215, 216
Safety Integrity Level, 159, 160, 200, 217, 218
Safety moment, 231
Safety Regulations, 242
Safety relief valve, 159
Safety screening report, 217
Safety Triangle, 230
Sand monitoring, 197
SAT, 200, 213, 284
Schedule, 20, 55, 87, 90, 117, 138
Secondary structures, 184
Security, 227, 233
Short circuit fault rating calculations, 213
Shutoff valves, 199
single line diagram, 208
Single source justification, 128, 284
Site Acceptance Testing, 200
site acceptance tests, 278
Site Queries, 269
Social Media, 42

Soft skills, xiii, xiv
Software for planning, 92
sports psychology, 77
Stage gate review, 158
Start-up, 205, 273
Stop card, 232
Stress, 79, 80, 81, 82, 83, 84, 85, 187, 188
Stressing the Business, 85
Stressors, 80
Structural analysis reports, 188
Structural integrity model holder, 188
Subcontract, 164
Subcontractors, 21
Supplier Reports Review, 130
Supply Chain, 124
Supply chain single point contact, 175
Supply vessels, 132
Surveillance and Witnessing, 260
System Architecture, 194
Systems, 138, 180, 192, 194, 197, 236, 237, 239, 242, 255, 271, 279
Team building, 30, 56
Technical and commercial bid evaluation, 128
Technical bid evaluation, 125, 127, 284
Telecommunication systems, 197
Temperature class, 212
Temporary Refuge, 215, 225
Terms &conditions, 126
Tertiary structures, 184
the Association for Project Management, xv
The Change Process, 135
Three-phase separation, 153
Three-point estimates, 121
Tie-Ins, 170
Time Management, 37, 38, 42
Toastmasters International, xv, 70, 286, 295
Trace heating, 211
training and coaching, 295
Transport and storage, 177
Triparty Walkdowns, 279
Twitter, 42
Types of Manager, 72
Uninterruptable power supplies, 207
UPS, 207, 208
Utilisation, 187
Value of Work Done, 111
variance narrative, 110
Vendors, 21
Verification, 189
Virtual Team Considerations, 54
Voltage transformers, 210, 211
Warehousing, 131
Waste management, 234
WBS, 93, 94, 105, 121, 265, 277
Weekly Worksheet, 40
Weight control reports, 189
weight limitations, 185

INDEX

Weld procedure qualification, 250
Weld type specification, 249
Welder coding, 249, 250
Welding, 238, 249, 284
Welding Procedure Specification, 249, 250
Work Breakdown Structure, 93, 277
Work- life balance, 76
Workforce fitness, 228
Workload, 48
WPS, 249, 250, 284
Yammer, 286

Peter F Cranston

About the Author.

Peter Cranston CEng, MIMechE, PMP, DTM has been a Project Engineer for over 30 years. After graduating from Edinburgh University with a BSc in Mechanical Engineering in 1981 he initially worked offshore as a remotely operated vehicle, ROV, pilot. His first real Project Engineering role was for a company building submarines for the tourist market which he describes as a "Boys own adventure in the Caribbean".

He has worked for main engineering contractors in a Projects role for the majority of his career on a variety of topsides, drilling facilities and subsea projects ranging from multiple modifications programs through to offshore modules and subsea tiebacks.

As a member of Toastmasters International for 11+ years he gained the highest award of DTM and as area governor managed 5 toastmasters clubs in Scotland. A portion of the communication and leadership skills developed at toastmasters has influenced the soft skills section of this book.

Peter enjoys his current role of training and coaching the new generations of project personnel and seeing both their skills and confidence improving. He is married to Gwendoline, has two grown up children and lives south of Aberdeen, Scotland.

Lightning Source UK Ltd.
Milton Keynes UK
UKHW052247090619
344110UK00004B/19/P